Saln

Data Warehousing in Construction Organizations

Salman Azhar

Data Warehousing in Construction Organizations

Concepts, Architecture and Implementation

VDM Verlag Dr. Müller

Imprint

Bibliographic information by the German National Library: The German National Library lists this publication at the German National Bibliography; detailed bibliographic information is available on the Internet at http://dnb.d-nb.de.

Any brand names and product names mentioned in this book are subject to trademark, brand or patent protection and are trademarks or registered trademarks of their respective holders. The use of brand names, product names, common names, trade names, product descriptions etc. even without a particular marking in this works is in no way to be construed to mean that such names may be regarded as unrestricted in respect of trademark and brand protection legislation and could thus be used by anyone.

Cover image: www.purestockx.com

Publisher:
VDM Verlag Dr. Müller Aktiengesellschaft & Co. KG
Dudweiler Landstr. 125 a, 66123 Saarbrücken, Germany
Phone +49 681 9100-698, Fax +49 681 9100-988, Email: info@vdm-verlag.de

Copyright © 2008 VDM Verlag Dr. Müller Aktiengesellschaft & Co. KG and licensors
All rights reserved. Saarbrücken 2008

Produced in USA and UK by:
Lightning Source Inc., La Vergne, Tennessee, USA
Lightning Source UK Ltd., Milton Keynes, UK

ISBN: 978-3-8364-3507-9

To my lovely wife Amna who is a continuous source of inspiration for me

TABLE OF CONTENTS

PREFACE

Construction organizations typically deal with large volumes of project data containing valuable information. It is found that these organizations do not use these data effectively for planning and decision-making. There are two reasons. First, the information systems in construction organizations are designed to support day-to-day construction operations. The data stored in these systems are often non-validated, non-integrated and are available in a format that makes it difficult for decision makers to use in order to make timely decisions. Second, the organizational structure and the IT infrastructure are often not compatible with the information systems thereby resulting in higher operational costs and lower productivity. These two issues have been investigated in this book with the objective of developing information systems that are structured for effective decision-making.

In this study, a framework has been developed to guide storage and retrieval of validated and integrated data for timely decision-making and to enable construction organizations to redesign their organizational structure and IT infrastructure matched with information system capabilities. The research was focused on construction owner organizations that were continuously involved in multiple construction projects. *Action research* and *Data warehousing* techniques were used to develop the framework.

One hundred and sixty-three construction owner organizations were surveyed in order to assess their data needs, data management practices and extent of use of information systems in planning and decision-making. For in-depth analysis, a public construction owner organization in Miami, Florida was selected. A functional model and a prototype system were developed to test the framework. The results revealed significant improvements in data management and decision-support operations that were examined through various qualitative (ease in data access, data quality, response time, productivity improvement, etc.) and quantitative (time savings and operational cost savings) measures. The research results were validated by a representative group of construction owner organizations involved in various types of construction projects.

The research presented in this book was conducted from January 2003 to December 2005. It is expected that this book will provide useful information to construction organizations considering data warehousing implementation for better decision-support.

I extend my gratitude and appreciation to the following people who directly and indirectly contributed to the structuring of this work: *Prof. Irtishad Ahmad, Dr. Syed M. Ahmed, Dr. Ronald Giachetti, Dr. Luis Prieto-Portar* and *Dr. L. David Shen* from Florida International University, Miami, Florida; *Mr. George Navarrete, Ms. Maria Cerna* and *Mr. Rafael R. Cutie* from Miami-Dade Transit (MDT), Miami, Florida; and *Mr. L. Crane Johnson*, Paris Technologies Inc., Pennsylvania. Last but not least, I would like to thank my parents. Without their patience, understanding, support, immense love, and most of all their countless prayers, the completion of this work would not have been possible.

Salman Azhar

Auburn, Alabama, USA

October 2007

CHAPTER 1

INTRODUCTION

1.1 Research Background

The project data in construction organizations are typically stored in operational and applications databases of information systems (IS) to support construction operations and decisions. The data are often non-validated, non-integrated and stored in a format that makes it difficult for the decision makers to make quick decisions. This is a typical problem in many business organizations and it is due to data modeling limitations of existing information systems, traditional organizational setup and inadequate information technology (IT) infrastructure (Ang and Teo, 2000). Moreover, in construction organizations, the databases are likely to be geographically and/or functionally dispersed, thereby making it difficult and inconvenient to access in a relatively short time available for decision-making. A decision maker may have to wait days or weeks for responses from subordinates that handle requested database queries in order to extract required information from the data. Such long waiting periods can have adverse impact on project performance and may reduce the value of information (Ahmad and Azhar, 2005).

The databases in most construction organizations, are based on the concept of Online Transaction Processing (OLTP). These OLTP databases are updated continually on a periodic basis and are suitable to support day-to-day business operations. Such transactional databases are designed to answer *who* and *what* type questions, they are not very effective in answering *what-if*, *why*, and *what-next* type queries (Ahmad and Azhar, 2002). The reason is that data in OLTP databases are not necessarily organized to support analytical processing and decision-making.

For example, payroll data can provide information on wages of regular and overtime labor. Additionally, scheduling database can provide information on project progress. However, if an estimator or scheduler is interested in knowing "what is the effect of employing overtime labor on project productivity?", conventional construction databases cannot provide a direct answer. The executive making this query must rely on subordinates to extract relevant data from accounting and scheduling databases. These data need to be interpreted, and then necessary computations are needed, in order to get the answers. The whole process could take days to weeks depending on the size and complexity of the project (Yang and Yau, 1996). The issue is to be able to quickly analyze existing data to discover trends so that predictions and forecasts can be made with reasonable accuracy and in a timely manner to aid in the decision-making process.

A number of research studies have been conducted with the objective to integrate transactional information systems and develop decision-support systems (DSS) to analyze a problem or situation more effectively and to support the decision-making process (Anumba, 2000). However, such DSS systems are not very successful due to several inherent flaws. Firstly, they are not completely separated from the transactional systems and the sharing of data slows down, or sometimes halts, both the transactional operation and the analysis process. Secondly, they are not designed to, or are limited in their ability to, generate trend analysis, discover patterns, and establish multidimensional data relationships. Thirdly, the data quality is poor because data are often non-validated. (Chau et al., 2003; Watson et al., 2001). Hence executives would still not be able to make critical decisions based on the "lessons learned" from completed projects unless some specially prepared data are provided to them.

The information management capabilities of an organization strongly influence its design (Nosek, 1989). Organizational design is considered a dynamic process. A number of research studies indicated that organizational design and the design of work processes is shaped by the amount and type of information required in a given environment and the organization's information processing capability (Galbraith, 1974; Tushman and Nadler, 1978). In construction organizations, the organizational structure is often rigid and usually not redesigned after implementing new or enhanced information systems. This situation often results in higher operational costs and lower productivity because the organizational structure and information systems are not compatible with each other.

The purpose of this research is to develop a framework to guide storage and retrieval of validated and integrated data for timely decision-making and to enable construction organizations to redesign their organizational structure and IT infrastructure matched with information system capabilities. The research study examines *data warehousing,* a fairly recent database management technique for its applicability and benefits that can be realized in the construction industry. Data warehousing is an improved approach for integrating data from multiple, often very large, distributed, heterogeneous databases and other information sources. It supports reorganization, integration, and analysis of data that enable users to access information quickly and accurately. Data warehousing is based on Online Analytical Processing (OLAP) concept as opposed to OLTP. The OLAP facilitates analysis of data using statistical and computational techniques.

The research hypothesis is, currently construction organizations do not effectively utilize project data for planning and decision-making due to transactional nature of their existing information systems, ineffective organizational structures and inadequate IT infrastructure. If properly implemented; the proposed decision-support framework based on the data warehousing technique will enable construction organizations to effectively store and utilize project data for decision-making. It will facilitate coordination, promote greater interaction and responsiveness in the construction processes and reduce the direct and operational costs related with information processing and decision-making.

1.2 Research Challenges

Construction is an industry bound by traditions, not necessarily by choice, but because of the ways organizations are structured and have worked, over the years and because of their dependence on age-old conventions, norms and rules. Implementation of concepts like data warehousing is a challenging one in construction, as it would have an impact on these conventions, norms and rules. Thus the research is focused on deployment of IT through innovative applications by developing new ways of knowledge acquisition and management. The research encompasses organizational as well as technical aspects of the challenge of implementing IT, in general, and data warehousing, in particular, in a traditional industry. The concept of data warehousing is simple, technically appealing but when taken practical issues into consideration, its implementation in construction organizations poses enormous challenges. These challenges are organizational, financial, and technical. In this research study, all three areas of challenges are addressed. The results of this research will enhance our capacity to respond through IT to new opportunities, and to reduce the lag time between concept and implementation.

1.3 Objectives and Scope

The main purpose of this research is to develop a decision-support framework based on data warehousing principle for construction organizations. The framework provides schematic procedures to design and implement data warehousing technology, and guidelines to redesign organizational structure and IT infrastructure.

The objectives of the research, in specific terms, are listed below:

(1) To assess current data management practices, and extent of use of information systems in planning and decision-making in construction organizations.

(2) To develop a multi-perspective enterprise model of construction organizations to replicate construction-specific business processes, data and information flow rules, decision processes and information requirements for these decisions.

(3) To develop a functional model and reference architecture of the data warehouse.

(4) To suggest modifications in existing organizational structures and IT infrastructure with an aim to optimize the planning and decision-making process.

Although the major outcome of this study, the principles of the proposed framework are applicable to any group of construction organizations, the scope of the research is limited to owner organizations for three reasons: (1) Owners have a wider perspective as they are involved with the project from inception to completion, and in most cases, are the ultimate users of the constructed facilities; (2) Owners are in control of project funds, the extent of this control, however, depends on the type of contractual arrangements, but owners are the ultimate stakeholders, as far as the overall investment in a project is concerned; and (3) Owner organizations (public agencies, in particular) data are easily available in public.

For the purpose of this research study one-time construction clients are not considered as owners. The study is focused on the type of owners that are continuously involved in construction projects. This type includes public agencies, such as the state department of transportations, county/city and state governments, transit agencies, port authorities, school boards, etc. Private corporations, such as major franchise owners, chain-store owners, energy/utility companies, airlines, etc. are included in this group.

1.4 Research Approach

To achieve the research objectives, as outlined above, a relatively new philosophy of research named *action research* is employed in this study. This philosophy of research is particularly suitable for research studies involving social and/or business organizations and their functions. In action research, the researcher reviews the existing situation, identifies the problem, gets involved in introducing some changes to improve the situation and, evaluates the effect of those changes (Naoum, 2001). A brief outline of the research methodology follows:

(1) Assessment of data management practices and level of utilization of information systems for planning and decision-making in construction owner organizations through a questionnaire survey;

(2) Organizational, functional, information and decision modeling of selected owner organization;

(3) Development of functional model and reference architecture using data warehousing principle. It involves development of data extraction, validation and integration schemes, multidimensional data models, and OLAP query techniques;

(4) Development of the decision-support framework showing different organizational setups and matching IT infrastructure to effectively utilize project data for decision-making; and

(5) Validation of the decision-support framework. Both qualitative and quantitative measures are used for validation. Qualitative measures include improvements in data access, data quality, productivity, response time, etc. while quantitative measures consist of time savings and operational cost savings.

1.5 Organization of the Book

The research presented in this book was conducted from January 2003 to December 2005. The remainder of this book is divided into seven chapters. A review of the available literature is presented in Chapter 2. It covers research efforts in the areas of database management, decision-support systems, data warehousing and organizational design. An examination of the database management systems and decision-support systems developed for the use of construction organizations is presented, the concept and advantages of data warehousing over traditional database approaches are pointed out, and the data warehousing design strategies are discussed. At the end of the chapter, different organizational design methods are discussed with their advantages and disadvantages.

In Chapter 3 the research design and detailed methodology are discussed. The reasons for adopting the action research technique are also outlined.

The design and results of the questionnaire survey are presented in Chapter 4. The multi-perspective modeling of the selected owner organization is illustrated in Chapter 5. In Chapter 6, first the development of the functional model and reference architecture using the data warehousing concept is presented. Then the prototype system is demonstrated and discussed in the later part of this chapter.

The proposed decision-support framework is presented in Chapter 7. A discussion of the changes to be expected in the decision-making process after the implementation of the framework and its impact on the organization's performance are discussed. The last two sections of this chapter illustrate the validation of the framework by selected owner organizations.

In Chapter 8, the research is summarized, conclusions are drawn, and the important findings are emphasized. In the last section of this chapter, directions for future research are outlined.

CHAPTER 2

DATA WAREHOUSING SYSTEMS IN CONSTRUCTION: AN OVERVIEW

2.1 Introduction

Use of computers for processing operational data and for providing information to decision-makers began about three decades ago. Nowadays, explosive growth of many businesses, scientific and government databases has far outpaced our ability to manage and interpret data. Traditional methods can create summarized reports from data, but cannot analyze the data explicitly to support decision-making. A critical need exists for a new generation of techniques and tools with the ability to automatically analyze the mountains of data for planning and decision-support (Soibelman, 2000).

In this chapter research efforts in database management, data warehousing and decision support systems in construction are reviewed. First presented is a brief review of database management systems and decision-support systems developed for the construction industry with their relative strengths and shortcomings. Then basic concept behind the data warehousing technology and its advantages over traditional databases are discussed with the help of several case studies. Information flow within an organization and availability of appropriate information for decision-making are greatly dependent on the setup or structure of that organization. With this focus, various organizational design processes are examined at the end of this chapter.

2.2 Database Management Systems in Construction

The use of database management systems (DBMS) in civil and construction engineering is not new. Like the service and the manufacturing industries, the construction industry has benefited from the advancements that took place in the DBMS technology. Earlier systems were developed for stand-alone applications such as the CAD systems, estimating, scheduling and inventory control (Chen et al., 1994). Efforts were made in the early 1980's to integrate various construction-related databases. Much of the early integration efforts revolved around the idea of integrating 2D drafting with 3D modeling, integration of graphical and non-graphical (or computational) design information, and integration of two or more applications, such as analysis and design (Amor and Anumba, 1999). In the 1990's, emphasis was placed on the process integration (i.e. design and construction) and the multidisciplinary integration (such as integrating various construction units). During that time, different interfaces were developed to integrate CAD systems with project planning, scheduling, cost estimating and cost control tools and methods. Examples of such efforts, available in the literature, include COMBINE®, CIMSTEEL®, ATLAS®, and ICON® among others (Bjork, 1999). Another important development was the formulation of standards for data exchange such as STEP® (Standard for the Exchange of Product Data), and IFCs™ (Industry Foundation Classes).

In the mid 1990's, the widespread use of internet and the World Wide Web (WWW) inspired the researchers to develop database applications based on internet and intranet technologies. In 1998, Jacobsen et al. presented database architecture for an integrated environment supporting team design, where people could work and communicate from different places. This was one of the early developments in the era of web-based project management systems. Abudayyeh et al. (2001) developed an intranet-based cost control system to facilitate cost control by improving the quality and timeliness of information. The system resulted in several benefits such as availability of instant, automated, on-line

10

reports produced on-demand, reduction in paperwork and better sharing of information among project participants.

Recently, with advances in computer software and hardware technology and due to exponential growth of Internet-based communication technology, the trend has shifted towards the development of online project management systems, nD visualization systems (CAD products integrated with construction process databases), and Enterprise Resource Planning systems (ERP). In the following paragraphs, some of the recent research efforts on these topics around the world are briefly discussed.

The AIC (Automation and Integration in Construction) Research Group at the University of Salford, UK has developed a web-based integrated environment prototype called SPACE® (Simultaneous Prototyping for An integrated Construction Environment). SPACE® provides users with a multi-disciplinary computer environment where project information can be shared between the various construction professionals using integrated databases (SPACE, 2002). The basic architecture of SPACE® is shown in Figure 2.1.

CIMS® (Computer Integrated Management System) developed at the Hong Kong Polytechnic University provides an integrated information control, material control, cost control, progress control, human resources management and tools for quality assurance. The architecture of CIMS® is presented in Figure 2.2 (CIMS, 2002).

There has been extensive research work done on the development of 4D CAD systems (3D + time) at academic institutions like Stanford University, VTT Finland, University of Loughborough, Hong Kong Polytechnic University, and at leading construction companies such as Bechtel Inc. and Flour Daniel Inc. The main objective of these research studies is to integrate 3D models developed using CAD softwares with schedules generated by project planning softwares (such as Primavera Project Planner™). The purpose is to enable project managers to create and update realistic schedules rapidly and to integrate the temporal aspects of a schedule as intelligent 4D models. Four dimensional (4D) models include powerful visualization, simulation and communication tools that provide simultaneous access to design and scheduling data, benefiting many project participants, not just the construction organizations (CIFE, 2002; VTT, 2002; BRE, 2002; Rischmoller and Alarcon, 2002). The workflow diagram illustrating the fundamental concept of 4D modeling is shown in Figure 2.3.

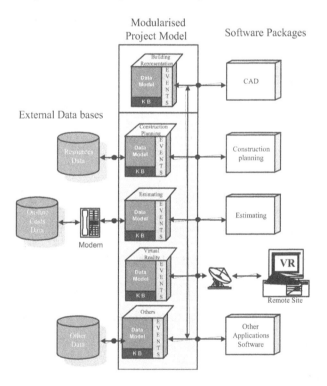

Figure 2.1: Architecture of SPACE® (SPACE, 2002)

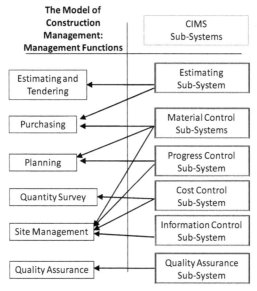

Figure 2.2: Computer Integrated Management System, CIMS® (CIMS, 2002)

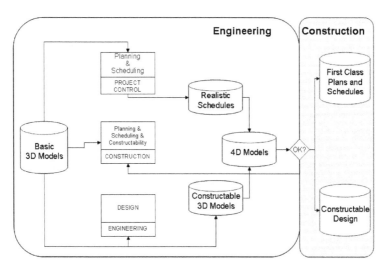

Figure 2.3: Simplified Workflow Diagram of 4D Modeling
(Rischmoller and Alarcon, 2002)

Enterprise Resource Planning (ERP) is another area that is now being hailed as a foundation for the integration of organization-wide information systems. Research efforts in this area have just begun and the development and implementation of ERP systems in the construction industry can be considered at its infancy. Shi and Halpin (2002) proposed a 3-tier architecture to implement ERP systems in construction organizations. The 3-tier architecture as shown in Figure 2.4 would assist construction enterprises in optimizing the utilization of their internal and external resources for maximizing the achievement of business objectives. A study by ML Payton Consultants (2002) indicated that more than half of the large US and European construction companies have some sort of ERP systems (customized/in-house or external/commercial) in place. The extent of implementation, however, varies and most are not very effectively implemented or utilized. At least half of these companies complain that commercially available ERP systems are hard to customize for an individual company's operations and this is one of the main reasons behind their slow adoption or ineffective utilization (Ahmed et al., 2003). The top ERP systems for construction include One World® by JD Edwards, SAP® and BAAN®.

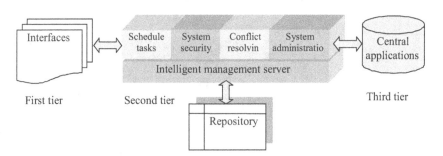

Figure 2.4: A 3-tier Architecture of Construct-ERP (Shi and Halpin, 2002)

14

The commercial software vendors are also making prominent progress towards the development of integrated project management softwares. One such example is Prolog Manager™ developed by the Meridian Project Systems Inc. Prolog Manager™ automates everyday management functions and provides features to track the project from design to close out. Built on the MS SQL Server™ 7.0 database platform, Prolog Manager™ enables multi-project control of procurement, cost controlling, document management, collaboration, and field management within a single application as shown in Figure 2.5 (Prolog Manager™, 2002).

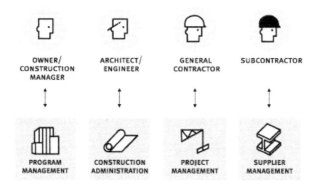

Figure 2.5: Different Features of Prolog Manager™ (Prolog Manager™, 2002)

Various applications and efforts in database management systems mentioned above have the objective of providing accurate information about an ongoing or completed project activity. However, most of them are not designed to, or are limited in their ability to, generate trend analysis, discover patterns, and aid executive-level decision-making (Chau et al., 2003; Watson et al., 2001; Polevoy, 1999; and Jacobsen et al., 1997). Hence executives would still not be able to make critical decisions based on the success and failure of previous projects unless some specially prepared data are provided to them.

2.3 Decision-Support Systems in Construction

Decision-support systems (DSS) are information systems that are designed to support complex decision-making and problem solving. These systems on one hand save the time and effort of experienced users in arriving at optimal decisions, while on the other hand, they help inexperienced practitioners to emulate the reasoning process of experts in solving problems in a specific domain (Ahmad, 1990; Songer et al., 1992).

The literature search provides various applications of DSS in civil engineering and construction management. Earlier systems were developed using *data/model management concept*. These systems used analytical or mathematical models and data to solve managerial problems ranging in complexity from financial decisions using simple spreadsheets to optimal site location using integer programming (Chen et al., 1994). Examples of such systems found in the literature include *DSS for Retaining Wall Management Application* (Chahine and Janson, 1987), *DSS for Modeling Bid/No-Bid Decision* Problem (Ahmad, 1990), *DSS for Contractors Prequalification* (Russell and Skibniewski, 1990), *Interactive DSS for Building Construction Scheduling* (Kahkonen, 1994), *Construction DSS for Delays Analysis* (Yates, 1993), and *DSS for Project Cost Control-Strategy and Planning* (Hastak, 1994).

Over the last decade, the making of many organizational decisions shifted from the individuals to small teams within an organization or large groups of individuals from multiple organizations. This change pushed the researchers to develop DSS for groups rather than individual users hence started the era of *Group Decision-Support Systems (GDSS) or Group Support Systems (GSS)*. Paulson and Kim (1999) developed a GDSS to support decentralized decision-making in project scheduling and control. Recently

Shen et al. (2004) presented a *GSS* for aiding value management decisions in construction organizations.

The rapid growth of internet motivated researchers to use web-based technologies in the development of DSS and GDSS. Tserng and Lin (2002) presented a web-based DSS to help contractors in selecting proper on-listed subcontractors by considering the trade-offs between risk and profit. Ng et al. (2003) demonstrated the application of web-based DSS for contractors' selection in traditional design-bid-build projects. Palaneeswaran and Kumaraswamy (2005) developed a web-based DSS for design-builder prequalification process.

Recently, the trend is shifted towards the development of *Knowledge-based Decision-Support Systems (KDSS)*. These systems are hybrid of Decision-Support Systems (DSS) and Expert Systems (ES) that help solve a broad range of organizational problems using data from current and previous projects. Kumaraswamy and Dissanayaka (2001) developed a KDSS to help clients engage in proper procurement processes by referring to good experiences collected from previous projects and the up-to-date knowledge developed in the field. Soibelman and Kim (2002) presented a data preparation system for construction knowledge generation through knowledge discovery in construction databases. They used their KDSS to identify the causes of construction activity delays in several projects.

Besides these developments, the practical use of DSS in the construction industry is very limited due to several reasons such as: (1) the lack of a strong database component due to limitations of transactional databases as discussed in section 2.2; (2) the data quality is poor due to non-validation of data, different data structures and storage formats of scattered databases. Hence construction managers do not greatly rely on the results of these applications; and (3) these applications are all developed by computer specialists in information centers after lengthy data analysis, but sometimes not all the requirements of construction managers are embodied sufficiently (Chau et al., 2003).

For effective decision-making, there is a need to develop integrated database management systems that can organize quality data from different databases over a period of time and automatically filter and transmit them to decision-support systems. Recently developed data warehousing technique is a response to the limitations imposed by the lack of effective connection between conventional database management systems and current decision-support systems (Ahmad and Azhar, 2005).

2.4 The Data Warehousing Concept

The traditional view of database management is based on the need for data to support transaction processing (therefore such systems are called Online Transaction Processing systems). Data warehousing evolved as an answer to the requirement of supporting analytical processing for informed decision-making. Thus the focus in data warehousing is analytical processing and not transaction processing (Summer and Ali, 1996).

The primary purpose of a data warehouse is to provide easy access to specially arranged data that can be analyzed through different analytical models and tools (Kimball et al., 1998). A data warehouse is typically a read-only dedicated database system created by integrating data from multiple databases and other information sources (Ahmad and Azhar, 2002). It differs from transaction systems in the following way (Gray and Watson, 1998):

(1) It covers a much longer time horizon (several years to decades) than do transaction systems.

(2) It includes multiple databases that have been processed so that the warehouse's data are subject oriented and defined uniformly (i.e., "clean prearranged data").

(3) It contains non-volatile data (i.e. read-only data) which are updated in planned periodic cycles, not frequently.

(4) It is optimized for answering complex queries from direct users (decision-makers) and applications.

The basic concept of a data warehouse is illustrated in Figure 2.6. A data warehouse is generally populated with data from two sources. The most frequent source is the periodic migration of data from Online Transaction Processing (OLTP) systems (i.e. operational databases such as financial and accounting, scheduling, cost, and human resources). The second source is externally purchased databases (such as lists of incomes and demographic information, and in the context of construction, material pricing data, cost indices, etc.) that can be linked to internal data. A data warehouse collects all data into one system, organizes the data for consistency and for easy interpretation, keeps "old" data for historical analysis, and makes access to, and use of data a simple task so that users can do it themselves without technical proficiency in data handling (Corey et al., 1998). In nutshell, a data warehouse is informational, not operational; analysis and decision are support-oriented, not transaction-processing oriented; and usually client/server-based, not legacy host-based (Gray and Watson, 1998).

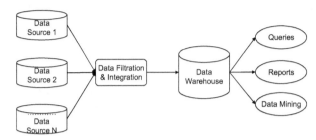

Figure 2.6: Conceptual Schema of a Data Warehouse (Adapted from Chau et al., 2003)

2.4.1 Data Warehouse Characteristics

A data warehouse is separate from the organization's operational or transactional databases. Table 2.1 summarizes the characteristics of a data warehouse.

Table 2.1: Characteristics of a Data Warehouse (Adapted from Mattison, 1996)

Characteristics	Description
Subject oriented	Data are organized around major subjects of the organization (e.g. productivity, finance) rather than individual transactions.
Validated	Inconsistencies are removed in both nomenclature and conflicting information. i.e. the data are clean and validated.
Nonvolatile	Read-only data; data are not updated by users.
Time series	Data are time series, not current status, i.e. a history of the subject over time, not a single moment in time.
Summarized	Operational data are aggregated, when appropriate, into forms suitable for decision-support.
Larger	Keeping a time series implies that much more data is retained. Size of data warehouse ranges from several gigabytes to terabytes.
Not normalized	Data can be redundant. (Normalization is a process of splitting a database into several files to avoid anomalies in storing, adding, updating, and processing).
Metadata	Data about the data for both users and data warehouse personnel. It contains information about the data stored in the warehouse.
Input	Operational data ("legacy systems") plus external data.

Another distinct characteristic of a data warehouse is the use of Online Analytical Processing (OLAP) concept for data analysis and decision-making. OLAP enable users to perform the following types of analysis over the data (Gray and Watson, 1998):

(1) *Categorical analysis*: Static analysis of data to generate data trends or patterns.

(2) *Exegetical analysis*: Drilldown capability to analyze a typical portion of the data in detail.

(3) *Contemplative analysis*: What-if analysis by changing a single variable and studying its effect on other variables.

(4) *Formulaic analysis*: What-if analysis by changing multiple variables and studying their effects.

2.4.2 Data Warehouse Architecture

Typically, the data warehouse architecture has three components or tiers, as follows (Vassiliadis et al., 2001):

(1) Data acquisition tools (back end) that extract data from transactional databases (i.e. OLTP systems) and external sources, consolidate and summarize the data, and load it into the data warehouse.

(2) The data warehouse itself contains the data and associated software for managing the data.

(3) The client (front end) software that enables users to access and analyze data in the warehouse.

The generic architecture of a data warehouse is illustrated in Figure 2.7. It can be seen from Figure 2.7 that data sources include existing operational databases and flat files (i.e. spreadsheets or text files) in combination with external databases. The data are extracted from the sources and then loaded into the data warehouse using various data loaders such as the SQL loader. The warehouse is then used to populate the various subject (or process) oriented *data marts* and *OLAP servers*. Data marts are subsets of a data warehouse categorized according to functional areas depending on the domain (problem area being addressed) and OLAP servers are software tools that help a user to prepare data for analysis, query processing, reporting and data mining. The entire data warehouse then forms an integrated system that can support various reporting and analysis requirements of the function of decision-making (Chaudhuri and Dayal, 1997).

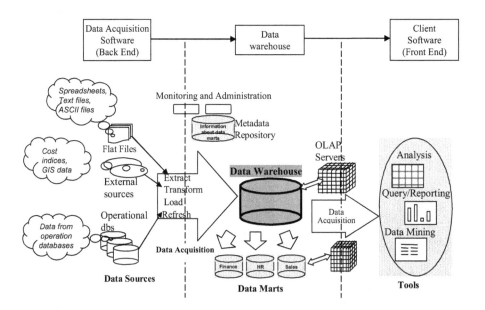

Figure 2.7: A Generic 3-Tier Data Warehouse Architecture
(Adapted from Chaudhuri and Dayal, 1997)

Data warehouse can also be implemented using *two-tier* and *one-tier* architectures. In two-tier architecture, the client software forms one tier while data warehouse and data acquisition software make the other tier. A stand-alone one-tier architecture, in which all three functions are on the same physical machine, is sometimes used when the amount of data is limited and the number of users is small (Gray and Watson, 1998).

The three-tier architecture is preferred due to its efficiency, flexibility and scalability (Dyché, 2000). It allows specialized databases to be distributed among various departments as data marts. Data in summarized form can be transmitted to relevant data marts during off-peak hours and made available for analysis during the day. Since the data warehouse forms a separate tier, it is easy to upgrade the software and hardware components. Two-tier architecture is suitable for medium-size organizations with fewer departments. It is economical and easy to implement. However the parallel data acquisition and data warehousing operations may slow down the performance of the entire system (Gray and Watson, 1998).

2.4.3 Data Warehouse Development Methodology

For transaction processing systems, the *Systems Development Life Cycle (SDLC)* is used to manage the development of applications. It breaks the development process into a set of interrelated steps. The typical steps include establishing the business needs, identifying information requirements, logical systems design, physical system design, programming, testing and implementation. Each step is required and feeds later steps. The purpose of SDLC is to bring structure and order to the development process. It requires that information requirements be determined early on. This approach forces users to think carefully about their requirements and discourages changes at a later stage.

Although the SDLC is well ingrained in IS practice for transaction systems, unfortunately it does not work well with data warehousing because it implicitly assumes that users know what they want. In data warehousing, typically users have some idea about their information requirements but only really find them out through a trial-and-error process. They need to see and use an initial version of the system and then specify the changes that need to be made (Vassiliadis et al., 2001). Hence the best approach for developing data warehousing systems is *Iterative Design* which sometimes called as *Evolutionary Design* or *Rapid Application Design (RAD)*. In this approach, a prototype of the system is developed based on initial user requirements and tested by different group of users. Based on their feedback, the prototype is modified and tested. This process is repeated until all the requirements are met and then the prototype is blown up into a full scale system (Dyché, 2000). The iterative design methodology steps are illustrated in Figure 2.8.

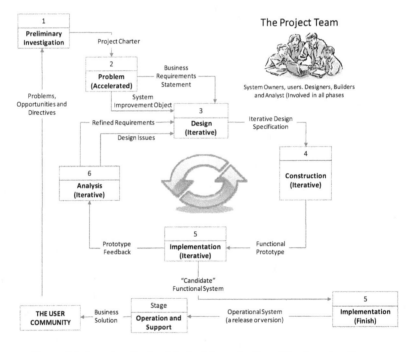

Figure 2.8: Iterative Design Methodology for Data Warehousing Systems
(Adapted from Whitten et al., 1998)

24

2.4.4 Data Warehouse Implementation

There are two basic approaches to implement a data warehouse, *Top-down approach* and *Bottom-up approach*. In the following sections, both approaches are discussed with their advantages and disadvantages.

2.4.4.1 Top-Down Approach

In Top-down approach, a full-scale enterprise-wide data warehouse is constructed prior to the introduction of smaller data marts as illustrated in Figure 2.9. This approach tacitly assumes that a data warehouse will contain all attributes needed for business applications and a data mart would comprise a subset (typically more summarized) of the data warehouse for a particular business application (Inmon and Hackathorn, 1994).

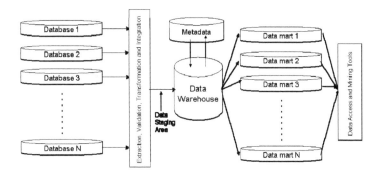

Figure 2.9: Top-Down Data Warehousing Implementation Approach
(Adapted from Fumo, 2003)

This approach requires complicated and time-consuming enterprise modeling which results in longer implementation times and higher costs (White, 2000).

2.4.4.2 Bottom-Up Approach

As introduced by Kimball (1997), it is an incremental approach to build data warehouse with significant cost savings and less implementation time. In this approach,

stand-alone data marts assigned to individual business units or processes are developed
and later integrated into an enterprise-wide data warehouse. The data staging area is
separate for each data mart as shown in Figure 2.10.

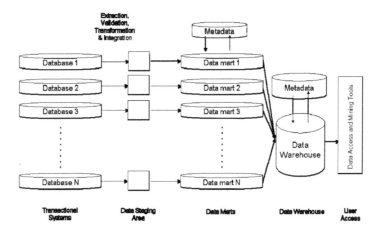

Figure 2.10: Bottom-Up Data Warehousing Implementation Approach
(Adapted from Fumo, 2003)

Both approaches have advantages as well as limitations. The Top-down approach
dictates that the central data warehouse is built first. All types of atomic and summarized
data are stored in the data warehouse and then subsets of data are created in the
distributed, subjected-oriented data marts. The problems with this approach are massive
effort and lengthy time required to define the enterprise data needs. This often causes
data warehousing efforts to stall and ultimately fail. The Bottom-up approach establishes
distributed, subject-oriented data marts first, using a multi-phase approach, and then
combines the data from all data marts into an enterprise-wide data warehouse. The
benefits of this approach are less development time, quick return on investment, and a
chance to leverage lessons learned as the team moves from one data mart to another. The

down-side of this approach is the chance of data inconsistencies among different data marts. It is very easy to make design or quality concessions within the individual data marts that make future integration into the enterprise-wide data warehouse very difficult (Inmon, 2000).

Domenico (2001) conducted a survey, with a focus of selecting the best approach, among different enterprises who implemented data warehousing. His research results conclude that Bottom-up approach is more successful due to the following reasons, (1) As it is incremental in nature, it allows more flexibility to change the data warehouse features according to user requirements; (2) It is easy to increase the scale of data warehouse with growing business needs and users demands; (3) It offers less risk as the users can start with few data marts which can later be grown into a full scale data warehouse.

Many researchers have suggested a parallel or hybrid approach by incorporating the advantages of both approaches (Barker, 2000; Watson, 2001; Domenico 2001; Shin 2002). This hybrid approach is illustrated in Figure 2.11. In this approach, planning phase is conducted according to Top-down approach however the implementation follows the Bottom-up methodology (Shin, 2002).

The hybrid approach offers two major advantages. First, the data marts could be implemented quickly and still highly beneficial to end-users who could not wait for completion of the enterprise-wide data warehouse. It allows rapid deployment of a decision-support application for a particular business unit that needs immediate attention. Second, this approach allows data mart activities to have a certain degree of independence from those of a data warehouse. Also, data mart project experience could render valuable insights concerning the incremental implementation of an enterprise-wide data warehouse.

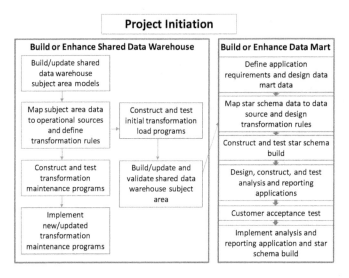

Figure 2.11: Hybrid Approach for Data Warehouse Implementation
(Shin, 2002)

The hybrid approach also solves the integration problem as all data marts share common business rules, semantics, and the enterprise metadata repository enforces consistent data definitions (Watson, 2001).

2.4.5 Decision-Support in the Data Warehouse

The fundamental role of a data warehouse is to provide data for decision-support. In some cases, the kinds of applications for which data warehouses are used have existed for some time (e.g., decision-support systems, executive information systems). However, data warehousing has added new life to them by improving and expanding the scope, accuracy, and accessibility of data. It has also led to development of new applications, such as data mining and knowledge data discovery (Krishna, 2004).

Online analytical processing (OLAP) is the heart of a data warehouse used for decision-support. OLAP is used for array-oriented applications such as market analysis, financial forecasting, etc. Besides these applications, OLAP should be used where (Gray and Watson, 1998):

(1) Requests for data are analytical, not transactional in nature.

(2) Significant calculations and aggregation of transaction-level data are involved.

(3) The primary type of data element being analyzed is numeric.

(4) Cross-sectional views of data are often required across multiple dimensions and along multiple consolidation paths.

The decision-support applications based on data warehousing and OLAP can be classified into four categories as follows (White, 2000):

(1) Queries and Managed Query Environment (MQE)

(2) Decision-Support Systems (DSS)

(3) Executive Information Systems (EIS)

(4) Data Mining (DMi)

2.4.5.1 Queries and Managed Query Environment

Queries refer to questions asked by an analyst from a database. If an organization does not have a separate data warehouse or data mart, decision-support analysts have to access the OLTP systems to obtain answers to their queries. Because many queries are large and complex, requiring a long time to execute, individual queries can interrupt the transactional systems for unacceptably long periods and even bring a system down. The data warehouse environment ameliorates this problem (Dyché, 2000).

Structured Query Language (SQL) is the standard for conducting database queries. SQL can be used in either a stand-alone or an embedded mode. In the former, the user writes SQL statements to access data. In embedded mode, the user is given a graphical front end that allows search conditions to be specified (Gray and Watson, 1998).

There are many potential problems with SQL; in particular, with stand-alone use. SQL requires the user to remember both SQL commands as well as names of data tables present in the database. This task is frequently error-prone, especially for more complex queries. Embedded SQL applications avoid these problems but queries are limited to those that are preplanned (Chau et al., 2003).

A Managed Query Environment (MQE) is designed to reduce or eliminate the above-mentioned problems. Users of an MQE operate through a graphical user interface and point-and-click objects that have been established for them. Each object has an underlying SQL statement(s). To execute a query, a user points-and-clicks on the objects that have been defined. The results can be displayed in a text or graphical mode.

An MQE provides security. The users are given access to those objects only that they are authorized to use. The MQE also allows IS department to control and schedule how queries are executed. Queries can be prioritized and defined to run at certain times. An MQE allows IS department to apply computer resources more effectively while making it easier for users to access and analyze data (Gray and Watson, 1998).

2.4.5.2 Decision-Support Systems (DSS)

Decision-support systems are designed to support specific decision-making tasks, such as providing a risk analysis for a proposed capital expenditure or scheduling material transportation activities. The OLAP and data warehousing have provided enhanced capabilities to access and analyze data for DSS applications thereby increasing their efficiency and performance. Some of these capabilities include *Support for multiple interfaces; Access to external databases; Integrated decision-support; and Multi-tier security* (Krishna, 2004).

2.4.5.3 Executive Information Systems (EIS)

An executive information system (EIS) is used by senior managers to access wide range of summarized information. Now-a-days, due to the use of data warehousing and OLAP, most EIS have embedded DSS capabilities. For example, some of the information provided by an EIS might be generated by a forecasting model, or a user may be able to perform OLAP analysis (White, 2000).

2.4.5.4 Data Mining (DMi)

Data mining refers to finding answers about an organization from the information in the data warehouse that the executive or the analyst had not thought to ask. Its objective is to identify valid, novel, potentially useful, and understandable patterns in data. Data mining is also known as *Knowledge Data Discovery (KDD)*. Data mining involves looking in the data for the following factors to solve the problem (Gray and Watson, 1998):

(1) *Associations*:   things done together

(2) *Sequences*:   events over time

(3) *Classifications*: pattern recognition (rules)

(4) *Clusters*:   define new groups

(5) *Forecasting*:   prediction from time series

Data mining can be *Bottom-up* (explore raw facts to find connections) or *Top-down* (search to test hypotheses). Therefore, data mining models can be developed using different algorithms such as Fuzzy logic; Neural networks and expert systems; Intelligent agents; Multidimensional analysis; Decision trees; etc.

2.5 Data Warehousing in Construction

In typical construction management information systems the database is designed to keep project records. Such systems consider construction processes as "temporary" and "specific", which means the data of one project can seldom be used for another project. Although construction projects are "unique", some similarities still exist between them, and construction processes and management skills are typically common to all projects. Hence, existing project data can be used for the planning and control of new projects (Chau et al., 2003).

During the project control phase, in order to take rectifying actions for any deviations in the performance, project managers often need timely analysis reports to measure and monitor construction performance. They also need timely analysis reports to assist in making long-term decisions (Vanegas and Chinowsky, 1996). It is found that most of the time is spent on collecting data from the various systems before the analysis can be made. Managers want and need more information, but analysts can provide only minimal information at a high cost within the desired time frames. In order to provide information for predicting patterns and trends more convincingly and for analyzing a problem or situation more efficiently, data warehousing along with its decision-support tools seems a promising solution (Ahmad and Nunoo, 1999; Lee and Lee, 2003).

Ahmad (2000) presented the concept of data warehousing in construction. This can be considered one of the first efforts in application of data warehousing in construction. In this paper, the author indicated data warehousing has a tremendous potential in construction. If implemented, data warehousing will facilitate a greater degree of coordination and will promote greater interaction and responsiveness in the process of construction. Data Warehousing will reduce the need for information processing and will eliminate unnecessary paperwork and bureaucracy. With data warehousing more information can be processed in a shorter period of time. Thus decisions will be made quickly, on a timely basis, and with appropriate information.

There are few recent instances of the application of data warehousing in construction. The following sections will review these applications and analyze their effectiveness with reference to decision-making in the construction industry.

2.5.1 Inventory Management

Chau et al. (2003) developed a Construction Management Decision-Support System (CMDSS) using data warehousing for inventory management. The system was designed for both novice and experienced users and was implemented at the Hong Kong Polytechnic University Student Dormitory construction project. The system could provide answers to complex queries such as "Determine the total amount of ceramic materials stored in the warehouses from suppliers in Beijing?" Using such queries, the construction managers could easily determine the inventory trend of the materials and the amount of each material type. It helped the managers in formulating an appropriate inventory decision or a warehouse storage strategy.

2.5.2 Accounting and Financial Management

A Chicago based construction company with offices located in nine states developed a data warehouse of its accounting and financial data. The purpose was to support the decision-makers for making critical price forecasting decisions and analyzing the company's profits and losses in different activities over a period of time. The system provides quick answers to executives for all types of forecasting and trend-type queries and generates ad-hoc reports whenever they are needed. Using a web interface, the company personnel have presentation-quality access to information at all hours. The system has resulted in substantial savings in preparing cost estimates for future projects and enhanced the productivity of the decision-makers (nSpin Case Study, 2002).

2.5.3 Site Selection for Residential Development

The site selection process depends on a number of spatial and business-related factors making it a complex decision-making task. It is common for the decision-makers to use their subjective judgment and gut feelings based on their experience in selecting the most appropriate site for development. The reason is that data for site selection originate from varied sources and are not organized in a format that decision-makers can readily use to derive any meaningful information. One possible solution of this problem is to develop a decision-support system (DSS) using data warehousing technique to help retrieve data from different databases and information sources and analyze them in order to provide useful and explicit information. Based on this concept, Ahmad et al. (2004) developed a DSS to assist builders/developers in site selection for residential housing development. A conceptual model of the developed DSS for site selection is shown in Figure 2.12. The DSS utilizes GIS application and employs the data warehousing technique in order to narrow a vast list of available sites for sale down to a manageable short list of a few

technically feasible sites. Subsequently, an analytical processing technique is used to

rank order the candidate sites for selecting the most appropriate one. The DSS can be

used for both short-term and long-term decision-making.

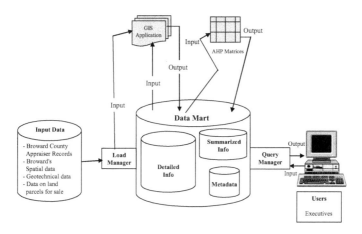

Figure 2.12: Conceptual Model of the DSS for Site Selection
(Ahmad et al., 2004)

2.5.4 Analyzing Project Documents for Decision-Support

Construction projects typically involve a large number of documents, which are

exchanged among multiple parties, including the owner, contractors and engineers.

Recently, the increased use of Web-based project management systems has made the

exchange of electronic documents a necessity in the construction industry. As a result, a

large number of structured electronic documents are accumulated in the database in each

construction project. To utilize the accumulated documents to support future decision-

making processes, Zhiliang et al. (2005) developed a system named EXPLYZER which

can extract useful information from the accumulated documents through the use of data warehousing technique and then analyze it. The system adopted a data standard based on the XML (Extensible Markup Language) to extract information from documents generated by a Web-based project management system. The conceptual model of EXPLYZER is presented in Figure 2.13. The EXPLYZER is composed of two modules, first module is for documents management and second module is for in-depth data analysis. Both modules are connected through a common graphical user interface (GUI).

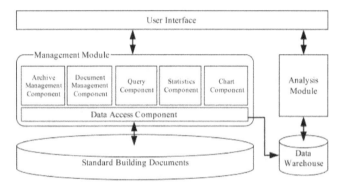

Figure 2.13: Conceptual Model of EXPLYZER® (Zhiliang et al., 2005)

The system can search across different kinds of documents to extract relevant information and perform various types of analysis such as the Earned value analysis, Productivity analysis, etc. based on user requirements.

2.6 Organizational Design for Effective Decision-Making

Organizations are designed to divide work among the interdependent subunits and coordinate the divided work in order to achieve the desired goals. There must be a continuous exchange of information and knowledge among the organizational subunits for effective decision-making (Mitropoulos and Tatum, 2000). The greater the task uncertainty, the greater the amount of information that must be processed among the decision-makers during task execution in order to achieve a desired level of performance (Galbraith, 1973).

As the task uncertainty increases and plans become less reliable, organizations have two primary ways to improve the information flow for decision mechanisms to select the best alternative course of action: (1) information systems; and (2) lateral relations.

In the past, information technology has supported routine decisions. Recent developments in information systems have increased user's ability to easily analyze less structured and more complex problems. For problems that are less quantifiable, group decision-support systems are proved to be very effective (Mitropoulos and Tatum, 2000).

Lateral processes may be informal, such as voluntary information exchange and cooperation for problem-solving, as well as formal. Super-ordinate goals, organizational environment, cooperative culture, and physical proximity of subunits facilitate informal cooperation. However, such processes do not always arise spontaneously; especially in highly differentiated organizations and more formalization is needed. Formal lateral processes include liaison roles, task forces, cross-functional teams and matrix organizations (Mitropoulos and Tatum, 2000).

When uncertainty and complexity increases, the decision-making process becomes more dependent on integration of information and knowledge among organizational subunits and external information sources (Jain, 1997). The different tools and techniques available for information integration have already been discussed in the earlier sections of this chapter; hence the focus in the following sections will be on organizational design and information requirements at different management levels for decision-making.

2.6.1 Organizational Design Process

Robey and Sales (1994) defined three approaches to the organizational design process, which are the *Emergent approach,* the *Proactive approach* and the *Reactive approach.*

2.6.1.1 The Emergent Approach

This approach considers organizational design as a dynamic process. It assumes that (1) the designer has only partial control over the design process, and (2) multiple participants influence organizational design. Figure 2.14 illustrates the nature of the emergent design process.

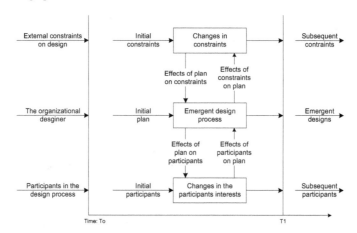

Figure 2.14: Characteristics of the Emergent Process of Organizational Design
(Robey and Sales, 1994)

As shown in Figure 2.14, the designer is in the middle of a dynamic process, surrounded by external constraints and a variety of participant groups. The designer's initial plans may have some effect on the constraints and participants, but the reverse is also partially true. The design process proceeds in this way over time, and subsequent designs reflect the influence of the designer, of other parties, and of external constraints. Because these factors interact and shift over time, the design process is best conceived as a fluid activity. The flows in and out of the design process are continuous and often unpredictable, which makes the management of the organizational design a challenging process. The emergent design approach is good for those organizations where complete managerial control over different processes cannot be exerted (Robey and Sales, 1994).

2.6.1.2 The Proactive Approach

The proactive or rational approach assumes that the organizational design is strongly shaped by the strategic plans and vision of the organization. This approach is based on the *Business Processes Reengineering (BPR)* concept. As different processes are reengineered, the organizational design is subsequently changed to reflect those changes (Mintzberg, 1993). This approach is good to design those organizations where a fundamental shift in organizational strategies has taken place. Figure 2.15 explains the proactive approach.

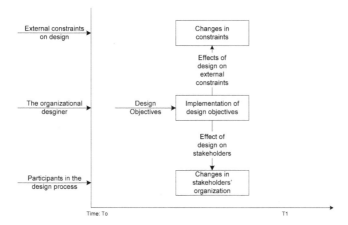

Figure 2.15: Characteristics of the Proactive Approach to Organizational Design
(Robey and Sales, 1994)

2.6.1.3 The Reactive Approach

The reactive or constrained approach assumes that the constraints determine the organizational design, with little or no chance for managers or other executives to produce alternative designs. This approach places heavy emphasis on constraints and depicts that the organizational designs that do not meet the demands of the constraints will simply not survive (Mintzberg, 1993). Figure 2.16 demonstrates the reactive approach.

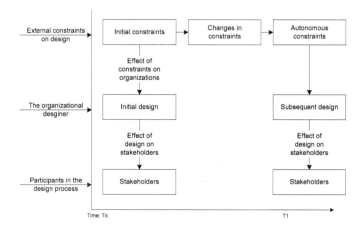

Figure 2.16: Characteristics of the Reactive Approach to Organizational Design
(Robey and Sales, 1994)

After analyzing all three design approaches, it can be argued that the proactive design approach is best suited to redesign construction organizations after implementation of new information systems due to three reasons. (1) It considers organizational design as a dynamic process; (2) It allows redesigning business processes after implementation of new information systems; and (3) It considers the effect of fundamental shift in organizational strategies. As discussed in chapter 1, the information management capabilities of an organization strongly influence its structure (Nosek, 1989). The proactive design approach provides an effective mechanism to change organizational structure in accordance with changes in the information processing capabilities of an organization.

2.6.2 Organizational Hierarchy and Level of Decision-Making

The need for information processing depends on the amount of information generated, number of organizational units involved, and interdependency among these units. Construction is an industry where vast amount of data are generated, processed and exchanged; where a large number of inter- and intra-organizational units are involved; and where the interactions among these units are often complex and sometimes confrontational. Therefore, construction organizations and their hierarchy should be structured in such a way as to satisfy these needs.

Each level of management in the hierarchy needs appropriate information in a suitable format. Management, depending on its level in the hierarchy, is responsible for making decisions aimed at assuring that the goals and objectives of the project under consideration are met (Kroenke and Hatch, 1994). These decisions may generally be categorized as Strategic, Implementation or Control, Functional or Operational, and Transactional. Executive management usually makes strategic decisions, whereas project management makes implementation decisions and functional management makes operational or functional decisions (Tang et al., 2004). In Figure 2.17, a typical organizational hierarchy with corresponding flow of information at different management levels is shown.

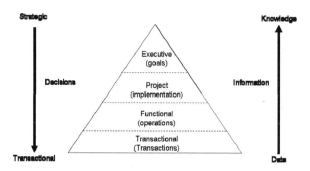

Figure 2.17: Flow of Information in a Construction Organization (Tang et al., 2004)

2.6.3 Organizational Design based on Information Processing Capabilities

Organizational design and the design of work processes are shaped by the amount and type of information required in a given environment and the organization's information processing capabilities (Galbraith, 1974; Tushman and Nadler, 1978). Watson et al. argued (2001) that since a data warehouse can provide more detailed, integrated, accessible, and historically complete information, it should be possible for an organization to operate very differently and "re-structure" itself.

Organizational design is the continuous monitoring and accessing of the fit between goals, structures, and rewards. Nosek (1989) considers "information and decision processes" as one of the key factors in organization design. His research concluded that organization design is a dynamic process and information management capabilities of an organization could strongly influence its culture and the structure. Kennedy (1994) compares the serial (vertical) and parallel (horizontal) organization structures to select the optimum structure based on the information processing capabilities of the organization. He found that the parallel structure provides better decision-support and also results in lower information processing costs.

Guha et al. (1997) investigated three companies that have implemented large scale information systems (IS). They found that the organizations that reshaped their design and strategies found tremendous returns. One of the companies, Lucent Corporation®, realized a US$250 million turnaround with six consecutive profitable quarters. On the other hand, a paper manufacturing company incurred only lower-level benefits, like reduced cost of operations and improved information access because it failed to redesign its structure and processes inline with the IS strategic objectives.

Dyché (2000) categorized the benefits of data warehousing in two categories namely *Soft Return on Investment* and *Hard Return on Investment* (ROI). Soft ROI included advantages such as customer satisfaction, increased productivity, technology leadership, cultural change and lower information processing costs while the hard ROI included benefits such as direct cost savings and increased market share. She pointed out that direct cost savings could be achieved by proper changes in the organizational design and strategies and could result in substantial monetary benefits.

2.7 Summary

The review of published literature revealed that there has been an active research in the areas of database management systems and decision-support systems for the construction industry from the past few decades. However, the research efforts focused on the development of systems to integrate various construction processes and to provide accurate information about an ongoing or completed project activity. These systems were not designed to, or were limited in their ability to generate trend analysis, discover patterns, and aid executive-level decision-making. Hence executives would still not be able to make critical decisions based on the success and failure of previous projects unless some specially prepared data is provided to them. Data warehousing technology can enable construction companies to consolidate data from diverse operational systems into one source for consistent and reliable information. The data stored in the data warehouse can be analyzed using different decision-support tools such as online analytical processing (OLAP) and data mining. The literature suggests that successful implementation of data warehousing also requires appropriate changes in the organizational structure and IT infrastructure.

CHAPTER 3

BUILDING THE DATA WAREHOUSE: STRATEGIES AND METHODOLOGY

3.1 Introduction

Due to multidisciplinary nature of this research problem, a thorough literature search

was conducted to identify various research methods employed in the field of information

systems, construction management and business management (Bouma and Atkinson,

1995; Creswell, 1994; Stewart and Michael, 1993; Yin, 1994; Zikmund, 1997). After a

careful analysis and mapping various research techniques with the research objectives,

the *action research* approach has been selected. Action research is an iterative technique

in which the researcher investigates the problem domain, identifies the problem, gets

involved in introducing some changes to improve the situation and evaluates the effects

of those changes (Naoum, 2001). Based on the action research technique, a five step

methodology has been designed which is depicted in Figure 3.1.

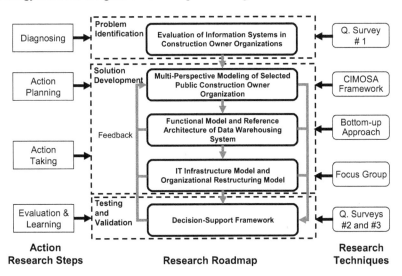

Figure 3.1: Research Methodology Framework

The necessary research data have been collected via questionnaire surveys, interviews, observations and project documents analysis. A focus group representing various functional units within the selected owner organization has been formed to get continuous feedback during each stage of the research. The research approach and methodology has been discussed in this chapter. First explained are the basic concepts and theoretical framework of action research. It is followed by brief explanation of different research steps designed within the framework of action research. Various business process and enterprise modeling techniques used at different stages of the research are also discussed.

3.2 The Action Research Approach

Action research is an established research method in use for scholarly investigations in information systems, construction management, business management, social and medical sciences. Action research aims at building/testing theory within context of solving an immediate practical problem in real setting. It combines theory and practice, researchers and practitioners, and intervention and reflection. The method produces highly reliable research results, because it is grounded in practical action, aimed at solving a realistic problem situation while carefully informing theory (Baskerville, 1999).

Hult and Lennung (1980) have defined the following three characteristics of action research which distinguishes it from other research methods:

(1) Action research aims at an increased understanding of an immediate problem situation, with emphasis on the complex and multivariate nature of organizations.

(2) Action research simultaneously assists in practical problem solving and expands scientific knowledge. This goal extends into two important process

characteristics: First, there are highly interpretive assumptions being made about observation; and second, the researcher intervenes in the problem setting.

(3) Action research is performed collaboratively and enhances the competencies of both researchers and practitioners.

3.2.1 Domain of Action Research

Action research addresses a specific problem situation, although it generates knowledge that enhances the development of general theory. The domain of action research method is characterized by a social setting where (Baskerville and Wood-Harper, 1996):

(1) The researcher is actively involved, with expected benefits for both researcher and organization.

(2) The knowledge obtained can be immediately applied, there is not the sense of detached observer, but that of an active participant wishing to utilize any new knowledge based on an explicit, clear conceptual framework.

(3) The researcher wants to link theory and practice to generate a solution.

One clear area of importance in the domain of action research is new or changed systems development methodologies. Studying new or changed methodologies implicitly involves the introduction of such changes, and is necessarily interventionist. From a social-organizational viewpoint, the study of a newly invented technique is impossible without intervening in some way to inject the new technique into the practitioner environment, i.e., "go into the world and try them out" (Wood-Harper, 1989). Action research is one of the few valid research approaches that can legitimately employed to

47

study the effects of specific alterations in systems development methodologies in human organizations (Baskerville and Wood-Harper, 1996).

3.2.2 Distinguishing Action Research from Consulting

Action research processes and typical organizational consulting processes contain substantial similarities. However action research and consulting differ in five key ways (Kubr, 1986; Lippitt and Lippit, 1978):

(1) Motivation: Action research is motivated by its scientific prospects, perhaps epitomized in scientific publications. Consulting is motivated by commercial benefits, including profits and additional stocks of proprietary knowledge about solutions to organizational problems.

(2) Commitment: Action research makes a commitment to the research community for the production of scientific knowledge, as well as to the client. In a consulting situation, the commitment is to the client alone.

(3) Approach: Collaboration is essential in action research because of its idiographic assumptions. Consulting typically values its "outsider's" unbiased viewpoint, providing an objective perspective on the organizational problems.

(4) Foundation for recommendations: In action research, this foundation is a theoretical framework. Consultants are expected to suggest solutions that, in their experience, proved successful in similar situations.

(5) Essence of the organizational understanding: In action research, organizational understanding is founded on practical success from iterative experimental changes in the organization. Typical consultation teams develop an understanding through their independent critical analysis of the problem situation.

In summary, consultants are usually paid to dictate experienced, reliable solutions based on their independent review. Action researchers act out of scientific interest to help the organization itself to learn by formulating a series of experimental solutions based on an evolving, untested theory (Baskerville, 1997).

3.2.3 Steps in Action Research Approach

Action research is a five phase cyclical process. The approach first requires the establishment of a client-system infrastructure or research environment (Susman and Evered, 1978). Then, five identifiable phases are iterated which are depicted in Figure 3.2:

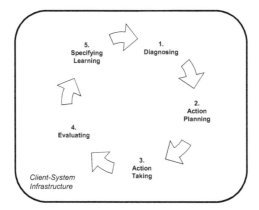

Figure 3.2: The Action Research Cycle (Adapted from Baskerville, 1999)

3.2.3.1 Client-System Environment

The client-system infrastructure is the specification and agreement that constitutes the research environment. It provides the authority, or sanctions, under which the researchers and host practitioners may specify actions. It also legitimates those actions with the express expectation that eventually these will prove beneficial to the client or host organization. Considerations found within the agreement may include the

49

boundaries of the research domain, and the entry and exit of the scientists. It may also patently recognize the latitude of the researchers to disseminate the learning that is gained in the research. A key aspect of the infrastructure is the collaborative nature of the undertaking. The research scientists work closely with practitioners who are located within the client-system. These individuals provide the subject system knowledge and insight necessary to understand the anomalies being studied (Clark, 1972).

3.2.3.2 Diagnosing

Diagnosing corresponds to the identification of the primary problems that are the underlying causes of the organization's desire for change. Diagnosing involves self-interpretation of the complex organizational problem, not through reduction and simplification, but rather in a holistic fashion. This diagnosis leads to develop certain theoretical assumptions (i.e., a working hypothesis) about the nature of the organization and its problem domain (Baskerville, 1999).

3.2.3.3 Action Planning

Researchers and practitioners then collaborate in the next activity, action planning. This activity specifies organizational actions that should relieve or improve these primary problems. The discovery of the planned actions is guided by the theoretical framework, which indicates both desired future state for the organization, and the changes that would achieve such a state. The plan establishes the target for change and the approach to change (Baskerville, 1999).

3.2.3.4 Action Taking

Action taking implements the planned action. The researchers and practitioners collaborate in the active intervention into the client organization, causing certain changes to be made. Several forms of intervention strategy can be adopted. For example, the intervention might be *directive*, in which the research "directs" the change, or *non-directive*, in which the change is sought indirectly. Intervention tactics can also be adopted, such as recruiting intelligent laypersons as change catalysts and pacemakers. The process can draw its steps from social psychology, e.g., engagement, unfreezing, learning and re-framing (Baskerville, 1999).

3.2.3.5 Evaluating

After the actions are completed, the collaborative researchers and practitioners evaluate the outcomes. Evaluation includes determining whether the theoretical effects of the action were realized, and whether these effects relieved the problems. Where the change was successful, the evaluation must critically question whether the action undertaken, among the myriad routine and non-routine organizational actions, was the sole cause of success. Where the change was unsuccessful, some framework for the next iteration of the action research cycle (including adjusting the hypotheses) should be established (Baskerville, 1999).

3.2.3.6 Specifying Learning

While the activity of specifying learning is formally undertaken last, it is usually an ongoing process. The knowledge gained in the action research (whether the action was successful or unsuccessful) can be directed to three audiences (Baskerville, 1999):

(1) The restructuring of organizational norms to reflect the new knowledge gained by the organization during the research.

(2) Where the change was unsuccessful, the additional knowledge may provide foundations for diagnosing in preparation for further action research interventions.

(3) The success or failure of the theoretical framework provides important knowledge to the scientific community for dealing with future research settings.

The action research cycle can continue, whether the action proved successful or not, to develop further knowledge about the organization and the validity of relevant theoretical frameworks. As a result of the studies, the organization thus learns more about its nature and environment, and the constellation of theoretical elements of the scientific community continues to benefit and evolve (Argyris and Schön, 1978).

3.3 Justification for Adopting Action Research Approach

Construction industry has been benefited by successful applications of action research in the areas of organizational modeling, strategic management, total quality management, information management, decision modeling, and risk analysis and mitigation (Naoum, 2001; Zikmund, 1998; Ahmad and Sein, 1997).

The action research approached has been adopted to conduct this study due to three reasons:

(1) The research identifies a real problem situation (i.e. ineffective utilization of project data in planning and decision-making) which is present in the construction industry from several years and yet no satisfactory solutions are developed.

(2) The research dictates the need of very close researcher-practitioner collaboration to find an adoptable solution. Construction is an industry bound by traditions, not necessarily by choice, but because of the ways organizations are setup, and have worked over the years and because of their dependence on age-old norms and rules. Implementation of new concepts is always challenging in construction, as it would have an effect on these set-ups, norms and rules. A close collaboration

with the construction industry practitioners will ensure that the new solution would be acceptable and applicable in the construction organizations.

(3) The research domain includes construction processes, information systems and organizational setup. Due to the different functionalities and behavior of these domains, the effectiveness of a solution can only be judged by its application and evaluation within the actual organization. The action research framework provides this functionality.

3.4 Selection of Owner Organization for Data Collection and Framework Testing

The action research methodology suggests that if the problem situation is present in a number of similar nature organizations then one organization may be chosen as a host for data collection, model/framework development, testing and validation. The research results then could be generalized for other organizations with some modifications (Azhar, 2007).

One public owner organization, Miami-Dade Transit (MDT) was chosen as a collaborator for this research study. This agency is in charge of administration of all public transportation-related construction projects in the Miami-Dade County. The reasons for this selection were the MDT's staff motivation about this research, firm commitment of executive management and grant of access to research-related data.

3.5 Research Methodology

As shown in Figure 3.1, the research was conducted in five interrelated steps. These steps are explained in the following sections. The different business process and enterprise modeling techniques utilized in this research are briefly explained. More details about these techniques are given in the relevant sections of Chapter 5 and 6.

53

3.5.1 Assessment of Information Systems in Construction Owner Organizations

Inline with problem diagnosing stage of action research, this step was designed to evaluate the data management practices in construction owner organizations and level of utilization of information systems in planning and decision-making. The questionnaire survey approach was selected to collect the necessary data. The target population was construction owner organizations which are continuously involved in multiple large scale construction projects. This includes both public as well as private owners. The questionnaire as shown in Appendix A was divided into three main sections as follows:

(1) Organization profile to elicit information about the respondent and characteristics of the surveyed organization.

(2) Assessment of data management techniques and degree of utilization of information systems in planning and decision-making.

(3) Evaluation of data warehousing need in the organization.

The questionnaire was sent to 550 selected owner organizations. The sample was drawn using stratified random sampling technique. Exploratory and inferential statistical methods were used to analyze the survey data. The questionnaire design, sampling criteria and analysis results are discussed in Chapter 4. The results of this survey are also used to validate the initial research hypothesis that construction owner organizations do not effectively manage and utilize project data for planning and decision-making.

3.5.2 Modeling of Selected Owner Organization

The purpose of owner organization modeling was to capture the current or "AS-IS" state of the organization and model it from various perspectives such as functional, organizational, informational, decision-making etc. These models were developed to understand the organization's business operations, flow of information within, into and outside the organization, identification of decision nodes and their hierarchy, and data requirements for these decisions. The information drawn from these models was used to recognize functional, informational and organizational requirements for data warehouse modeling.

The literature search indicated many standard enterprise modeling frameworks developed for the industrial systems but no such established method was found for the construction industry (Mak, 2001; Aouad, 1995; Howard, 1991). Hence it was decided to adopt one of the existing enterprise modeling framework and modify it to accommodate construction-specific requirements.

Among the available enterprise modeling frameworks, the *Computer Integrated Manufacturing Open Systems Architecture* (CIMOSA) was selected due to its simplicity, flexibility and worldwide acceptance. CIMOSA provides a process oriented modeling concept that captures both the process functionality and its behavior. It categorizes enterprise operations into *Generic* and *Specific* (Partial and Particular) functions. Generic functions are performed in every enterprise independent of its size, organization and business area. Examples include: control of work flow, administration of information, integration of resources and management of communications. Generic functions are specific to a particular organization only (Kosanke et al., 1999).

Figure 3.3 represents the CIMOSA framework (or reference architecture) in the form of a CIMOSA cube.

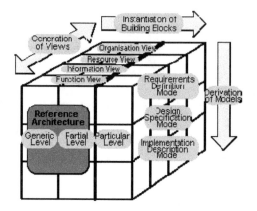

Figure 3.3: The CIMOSA Cube (Giachetti, 2002)

CIMOSA defines four modeling views/perspectives of the enterprise functions:

(1) Function view describes the work flows.

(2) Information view depicts the inputs and outputs of functions.

(3) Organization view indicates authorities and responsibilities of personnel.

(4) Resource view illustrates the structure of resources (humans, machines, etc.).

The Resource view is more appropriate for manufacturing enterprises which manage the supply chain of resources (Vernadat, 1993). Construction owner organizations usually do not involve in procurement operations under most types of contracts and hence this perspective is not useful for their modeling. Instead of resource perspective, a *Decision Perspective* was introduced to model the decision-making processes in the organization at different management levels.

Figure 3.4 illustrates the various steps undertaken during multi-perspective owner organization modeling. These steps are briefly explained in the following sections.

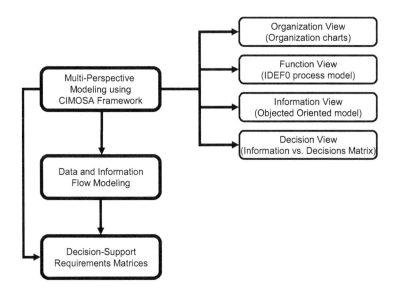

Figure 3.4: Steps Undertaken during Owner Organization Modeling

3.5.2.1 Multi-Perspective Modeling using CIMOSA Framework

The CIMOSA framework was used to model organization, function, information, and decision perspectives of the selected owner organization (i.e. MDT). Organizational hierarchical relationships were identified using organizational charts. For function modeling, IDEF0 (Integrated Definition for Function Modeling) technique was adopted. Information modeling was performed using Object Oriented modeling concept. For decision modeling, a matrix of construction decisions and respective information needs was prepared. The purpose of this matrix was to identify various construction-related decisions taken at different management levels and information requirements for these decisions. More details about these techniques are given in Chapter 5.

3.5.2.2 Data and Information Flow Modeling

In this step, operational data and their flow patterns in the organization were investigated by preparing data and information flow models. The purpose was, (1) to identify operational data and information flow into, within and outside the organization; (2) to examine degree of information processing at different management levels; and (3) to explore various detailed, summarized and exceptional reports prepared for different management positions.

3.5.2.3 Development of Decision-Support Requirements Matrices

In this step, decision-support requirements matrices for various management levels were developed by mapping data and information flow models with the decision model (i.e. construction decisions versus information needs matrix). These matrices were used to compare information desired and information available at various management levels for planning and decision-making operations. They also served as the guide map to capture user requirements for preparing the data warehousing functional model.

3.5.3 Development of Functional Model and Reference Architecture of Data Warehousing System

Based on the decision-support matrices and user requirements, a functional (or logical) model of the proposed data warehouse and its corresponding reference architecture were prepared. The functional model represents how the system would work while the reference architecture indicates its physical and technical implementation using different software and hardware tools. Figure 3.5 illustrates the framework adopted for developing the functional model and reference architecture. The functional model was developed using the Bottom-up approach as explained in section 2.4.4.1. As per this approach, first the data marts of different functional areas were developed and then integrated into the enterprise-wide data warehouse (Kimball, 1997). The reference architecture was prepared using 3-tier approach as explained in section 2.4.2. Based on the reference architecture, a prototype system was also developed. This step of research is explained in detail in Chapter 6.

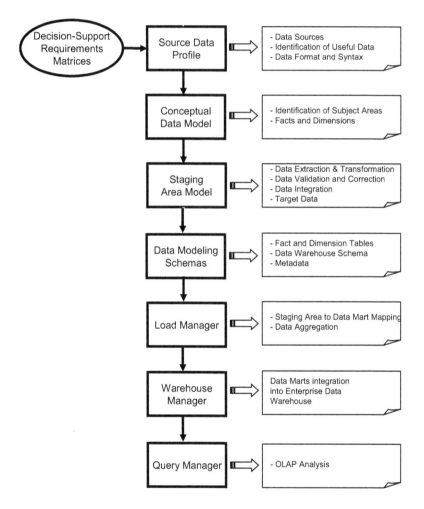

Figure 3.5: Framework Used for Preparing Functional Model and Reference Architecture

3.5.4 IT Infrastructure Model, Organizational Restructuring Model and the Decision-Support Framework

The decision-support framework provides schematic procedures to design and implement data warehousing technology in construction owner organizations, and guidelines to redesign their organizational structure and IT infrastructure. As discussed in chapter 1, the successful implementation of data warehousing requires adequate changes in the organizational structure and IT infrastructure. During this step, several new organizational structures compatible with the data warehousing system were prepared, discussed with the focus group of MDT, and tested using hypothetical but real project scenarios. The optimal organizational design was selected after several brain storming sessions and group discussions with the focus group. The new IT infrastructure was also developed to match the data warehousing implementation requirements. In the final form, the decision-support framework consists of data warehousing development and implementation schemes, new organizational setup and required IT infrastructure. The framework represents a modular architecture which makes it possible to modify it for different categories of construction owner organizations.

3.5.5 Validation of the Decision-Support Framework

The decision-support framework was validated using several quantitative and qualitative measures to ensure that the research objectives have been achieved. The validation procedure was conducted in two phases. In the first phase, the framework was validated for the Miami-Dade Transit (MDT) while in the second phase; it was validated for other construction owner organizations. The questionnaire survey approach was adopted for validation of the framework in both phases.

For validation at MDT, the focus group members were given detailed briefing about different functions of the framework, its hypothetical implementation in the organization, and demonstration of different decision-support situations with and without the framework. Their feedback was collected in two stages. The first stage represented the current or "AS-IS" state of the organization without any technological or organizational change. The second stage indicated that the proposed decision-support framework has been implemented in the organization. At each stage, the focus group members' feedback was collected through a questionnaire survey. The results were statistically analyzed to detect any significant improvements in various data management and decision-support practices and corresponding time and cost savings. The design of the questionnaire, analysis of data and comparison of results at both stages are discussed in section 7.5.

In the second phase, twenty construction owner organizations engaged in different types of construction operations were chosen to validate the framework. These organizations were chosen from a list of preselected companies who showed an interest in data warehousing implementation during questionnaire survey 1. Simple random sampling was performed to select a sample using 90% confidence level and 5% confidence interval. The framework validation was performed in two stages namely without and with framework implementation. The data collection and analysis was done in a similar way as phase 1.

The validation results of phase two justified that the proposed decision-support framework is applicable for all construction owner organizations with little or no modifications.

3.6 Summary

The research methodology is based on the action research approach. Action research is a novel approach for building and testing theory within context of solving an immediate practical problem in real setting. The research problem was diagnosed through a questionnaire survey targeted at different construction owner organizations. Miami-Dade Transit (MDT) was chosen as a collaborator for data collection and developed framework testing and validation. A modified version of Computer Integrated Manufacturing Open Systems Architecture (CIMOSA) was used to model organization using function, information, organization and decision perspectives. The end result of this modeling was a decision-support matrix which was used to capture organization and user requirements for the data warehousing system. The data warehouse was developed using a sequential Bottom-up approach and 3-tier architecture. A new organizational setup and IT Infrastructure compatible with the data warehousing system were proposed. The final decision-support framework consisted of data warehousing development and implementation model, new organizational structure and compatible IT infrastructure. At the end, the decision-support framework was validated first by MDT and then by twenty other construction owner organizations. In the validation step, organization's performance was measured before and after the implementation of the proposed decision-support framework using various qualitative and quantitative measures.

The various stages of decision-support framework development are discussed in detail in the following three chapters.

CHAPTER 4

ASSESSMENT OF INFORMATION SYSTEMS IN CONSTRUCTION OWNER ORGANIZATIONS

4.1 Introduction

The research need was established on a hypothesis that construction owner organizations do not effectively utilize project data for planning and decision-making due to lack of decision-support in their existing information systems. To validate the hypothesis, a questionnaire survey was conducted. Target population was public and private construction owner organizations which are continuously involved in multiple construction projects. The questionnaire was sent to 550 owner organizations and responses were collected in a period of three months. Statistical analysis is performed to analyze the survey data. The design of the questionnaire survey, its sampling, surveying method and major findings are discussed in this chapter.

4.2 Questionnaire Survey

4.2.1 Design

The questionnaire comprised of 15 questions which were grouped into 3 main sections namely (1) Organization profile; (2) Evaluation of data management practices and degree of utilization of information systems in planning and decision-making operations; and (3) Need of data warehousing in the surveyed organizations. A fourth optional section was provided at the end to collect personal information about the respondent. The complete questionnaire is available in Appendix A.

The purpose of the first section was to elicit information about the respondent and the organization itself. This section contained 6 closed-form questions inquiring about organization type, respondent's job function within the organization, nature of construction projects organization deals with, number of projects in the last five years, number of employees and annual construction expenditures. This information was later utilized to correlate organization characteristics with data management and decision-support practices in use.

Section two was designed to assess data management techniques adopted in the organization and degree of utilization of information systems in planning and decision-making operations. The first two questions in this section were closed form and were intended to collect information about the data management practices and database management systems (DBMS) in use. The next four questions were designed to gage respondent's opinion about the performance of their existing DBMS or information systems (IS) and level of their use in planning and decision-making operations. A 5-point Likert scale was adopted to determine respondent's degree of satisfaction (or agreement) and dissatisfaction (or disagreement) about a particular statement.

The aim of the third section was to evaluate the need of data warehousing in the organization. Before the start of this section, a brief explanation was provided about this technique to inform respondents who were unfamiliar with data warehousing. Three closed-form questions were designed to examine respondent's view about utilization of data warehousing technique for planning and decision-support and possibility of its implementation in their organization.

Fourth section was optional and respondent could voluntary provide his/her contact information. This information was later utilized to contact selected individuals and gather further information and explanation about their responses.

After designing the questionnaire survey, a pilot study was undertaken to measure the compatibility of questions with the objectives of the study, identification of any ambiguous question(s) or statement(s) and length of the survey. In the pilot study, the questionnaire was sent to a sub-sample of 15 individuals in 5 different owner organizations. These individuals were already aware about the study and their feedback was utilized to improve the questionnaire design.

4.2.2 Sampling

The target population of the survey was construction owner organizations which are continuously involved in multiple construction projects from the last ten years or more. This includes both public and private owners. Internet resources and yellow pages were used to identify such organizations. Table 4.1 illustrates ten major categories of construction owner organizations which were targeted for the survey.

Table 4.1: Targeted Categories of Construction Owner Organizations

1.	Departments of Transportation (DOT)	6.	School Boards
2.	Port Authorities (PA)	7.	Oil and Gas Companies
3.	Public Works Departments (PWD)	8.	Financial and Banking Corporations
4.	Mass Transit Agencies (MTA)	9.	Stores and Restaurant Chains
5.	Power Production/Distribution Companies	10.	Miscellaneous Public Organizations

Stratified random sampling technique was used to select sample for each category. In stratified random sampling, the entire population is divided into homogenous subgroups and then simple random sample is selected from each subgroup (Trochim, 2001). The criteria adopted for forming subgroups were nature of the organization and geographical regions. The reasons behind these criteria were to ensure that samples in each category have similar characteristics and are uniformly distributed across the continental United States. The sample size was determined using 90% confidence level and 5% confidence

65

interval. Table 4.2 illustrates the population size, sample size and response rate for each

category.

Table 4.2: Survey Population, Sample Size and Response Rate

No.	Category	Population (N)	Sample (n)	Responses Received	Response Rate (%)
1.	Departments of Transportation (DOTs) - Highways division - Aviation division - Railroads division - Other construction-related divisions	210	130	34	26.2
2.	Port Authorities (PAs)	155	100	28	28.0
3.	Public Works Departments (PWDs)	500	175	59	33.7
4.	Mass Transit Agencies (MTA)	35	30	12	40.0
5.	Power Production/Distribution Companies	35	30	9	30.0
6.	School Boards	24	20	6	30.0
7.	Oil and Gas Companies	15	15	4	26.7
8.	Financial and Banking Corporations	15	15	3	20.0
9.	Stores and Restaurant Chains	25	25	5	20.0
10.	Miscellaneous Public Organizations	10	10	3	30.0
	Total	1024	550	163	29.7

4.2.3 Method of Survey

The questionnaire was prepared in two formats: (1) Online or web-based format; and

(2) Printed or hard copy format. The majority of targeted organizations have their web

pages. This helped to locate the most suitable person in each organization who could best

answer the questionnaire. The personalized e-mails (or occasionally printed copies of the

cover letter and questionnaire when e-mail of a specific person was not available) were

sent to all identified persons explaining the purpose and objectives of the study,

importance of their feedback, instructions and web address (URL) of the survey. Such an

approach ensured that a high response rate is obtained. Importantly, respondents were

invited to actively participate in the work by providing constructive comments.

Reminders were sent after 30 days and 60 days of sending the questionnaire. Of the 550

questionnaires sent, 163 valid responses were received; representing a total response rate of approximately 30%. This response rate is typical of a construction industry questionnaire survey and can be used to draw any meaningful and unbiased conclusions (Akintoye and Macleod, 1997).

4.3 Findings

The data obtained from the questionnaire survey were statistically analyzed and major findings are reported in this section. Inline with the format of the questionnaire, the results are reported in three sections as follows:

4.3.1 Sample Characteristics

4.3.1.1 Organization Type

The owner organizations were broadly divided into two types, i.e. public agencies or departments and private companies. Figure 4.1 provides a breakdown of the valid responses by organization type. It indicates that 144 (88%) respondents were public agencies and 19 (12%) respondents were private companies. Such a breakdown was expected as the majority of construction owner organizations involved in multiple large-scale construction projects are public agencies.

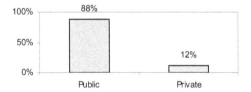

Figure 4.1: Respondents by Organization Type

4.3.1.2 Respondent's Main Job Function

Figure 4.2 delineates the respondent's main job function within the organization. It indicates that 82% respondents hold top management positions within the organization such as President or Vice President (17%), Construction chief (28%), Senior Manager (23%) and Project Manager (14%). On the basis of their position and work experience, it can be inferred that the respondents have adequate construction-related knowledge and are actively involved in planning and decision-making within their organization.

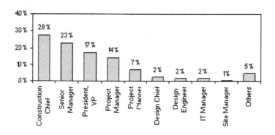

Figure 4.2: Respondents' Main Job Function

4.3.1.3 Types and Number of Construction Projects

The responding organizations had undertaken all types of construction projects ranging from infrastructure development and public works to residential and commercial projects. Each organization was typically found to be engaged in two or three different types of construction projects depending on the nature of the organization. Figure 4.3 illustrates the number of organizations involved with each type of construction project.

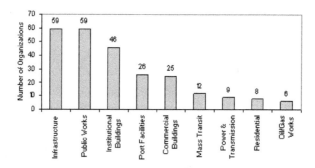

Figure 4.3: Types of Construction Projects Undertaken by Respondent Organizations

Figure 4.4 shows the breakdown of number of projects undertaken by each organization within the last 5 years. It indicates that around two-third organizations (69%) undertook 50 or more projects during the last five years. However, the number of undertaken construction projects also depends on the nature of organization and scope of the project. Organizations involved with mega-projects such as port development, mass transit etc. were found to be engaged in few large-scale projects while public works and private organizations were involved in many small-scale projects.

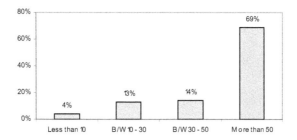

Figure 4.4: Number of Projects Undertaken by Organizations within the Last 5 Years

In nutshell, this information points out that the responding organizations are involved in all types and sizes of construction projects and the data collected from them comprehensively represents the construction industry as a whole.

4.3.1.4 Organization Size and Annual Construction Expenditures

Figures 4.5 and 4.6 provide details about the respondent organizations in terms of their number of employees (technical staff) and the annual construction expenditures. Of the 163 organizations surveyed, 65% have 500 or fewer employees while 35% have 500 or more technical personnel. The later category comprised of departments of transportation, port authorities and oil/gas companies. In terms of annual construction expenditures, there was no unique trend. Among the four categories of annual construction expenditures ranging from less than 25 million dollars to more than 100 million dollars, each category got around one-fourth (i.e. close to 25%) of the total share. This indicates that the survey sample consisted of small, medium as well as large size construction owner organizations.

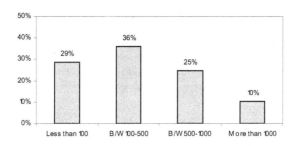

Figure 4.5: Organization Size by Number of Employees

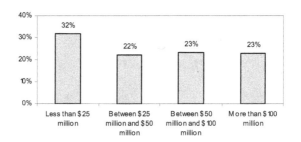

Figure 4.6: Annual Construction Expenditures of Organizations Surveyed

4.3.2 Evaluation of Data Management and Decision-Support Practices

4.3.2.1 Data Storage and Management Practices

The survey divided the data storage and management practices into two broad categories, *Manual* and *Computerized*. Manual means that the organizations store and manage data (including computer printouts) in paper files and folders while the term computerized indicates that the data storage and management is done through a formal

database management system (DBMS) such as Oracle™, Access™, SAP™, etc.

Tables 4.3 and 4.4 provide a summary of responses with reference to organization

types and sizes. Cumulatively, 48% organizations manage 50% of their data manually

and 50% via computerized means. Thirty-one percent organizations rely more on

computerized methods while 21% organizations were found to be using the traditional

manual storage techniques.

Table 4.3: Data Storage and Management Practices in Different Types of Organizations

Organization Type	Data Storage and Management (Manual/Computerized)				
	0/100	25/75	50/50	75/25	100/0
Departments of Transportation	3 (2%)	13 (8%)	15 (9%)	3 (2%)	
Port Authorities	1 (½%)	13 (8%)	10 (6%)	4 (2½%)	
Public Works Departments		10 (6½%)	27 (17%)	20 (12%)	2 (1%)
Mass Transit Agencies	1 (½%)	4 (2½%)	7 (4%)		
Power Companies		2 (1%)	6 (4%)	1 (½%)	
School Boards		1 (½%)	2 (1%)	4 (2½%)	
Oil and Gas Companies			3 (2%)	1 (½%)	
Financial and Banking Corporations			3 (2%)		
Stores and Restaurant Chains		1 (½%)	4 (2½%)		
Miscellaneous Public Organizations		2 (1%)	1 (½%)		
Total	5 (3%)	45 (28%)	78(48%)	33 (20%)	2 (1%)

Table 4.4: Data Storage and Management Practices v. Organization Size

Organization Size (No. of employees)	Data Storage and Management (Manual/Computerized)				
	0/100	25/75	50/50	75/25	100/0
Less than 100			17 (10%)	28 (17%)	2 (1%)
B/w 100 to 500		9 (6%)	45 (28%)	5 (3%)	
B/w 500 to 1000	2 (1%)	28 (17%)	10 (6%)		
More than 1000	3 (2%)	8 (5%)	6 (4%)		
Total	5 (3%)	45 (28%)	78 (48%)	33 (20%)	2 (1%)

The *One Way Analysis of Variance* test (ANOVA) was undertaken to statistically

determine if any significant differences exist between the type of organization, size of

organization and the data storage and management methods. ANOVA is normally used

71

to compare the means of more than two independent groups (Barnes, 1998). In this test, the hypotheses for the comparison of independent groups are:

H_o: Means of all groups are equal (no significant differences exist)

H_a: Means of the two or more groups are not equal (significant differences exist)

The significance level set to test the hypothesis was 5%, i.e. if $p<0.05$ it means there is evidence to reject the null hypothesis in favor of the alternative hypothesis. The results of the ANOVA test are summarized in Table 4.5. The results reveal that the data storage and management methods vary significantly with the type of organization and size of organization.

Table 4.5: Summary of ANOVA Test Results

Reference		Sum of Squares	df	Mean Square	F	p
Organization type	Between Groups	13.657	9	6.892	22.762	0.028
	Within Groups	1.201	40	0.300		
	Total	14.857	49			
Organization size	Between Groups	10.057	3	5.029	25.143	0.005
	Within Groups	0.800	16	0.200		
	Total	10.857	19			

An analysis of Table 4.3 shows that public agencies which are mostly involved in the development of large-scale infrastructure-related projects such as departments of transportation, port authorities and transit agencies mostly use computerized means to storage and manage project data. The public works departments are found to be more inclined towards manual storage methods. Tables 4.4 and 4.5 further support this finding and points out that large-size organizations (with 500 or more employees) use computerized methods while small-size organizations (with 500 or less employees) employ manual techniques.

72

4.3.2.2 Types of Database Management Systems in Use

Figure 4.7 illustrates the types of database management systems (DBMS) used by the surveyed organizations. It indicates that the most popular choice is the Spreadsheet systems (e.g. Excel™, Lotus™, etc.) which are used by approximately 90% organizations to store and manage data. The obvious reasons behind this selection are their low-cost, user-friendliness and easy availability. Besides using Spreadsheet systems, most organizations also use another dedicated database management system to archive the projects data. Around 40% organizations indicated that they use MS Access™, 30% use Oracle™, 11% use SAP™ (ERP based software), 10% use FoxPro™ or DB4™ and the remaining 9% organizations use in-house developed systems for data storage and management.

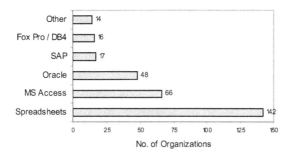

Figure 4.7: Types of Database Management Systems in Use

4.3.2.3 Level of Satisfaction with the Current Database Management System

The respondents were asked to indicate their level of satisfaction with the current database management system (DBMS) using six predefined satisfaction criteria. Their responses are summarized in Table 4.6. A quick look at the mean and standard deviation

73

values indicate that most of the respondents were not satisfied with their current database management system.

Table 4.6: Level of Satisfaction with the Current Database Management System

Satisfaction Criteria	Mean	Standard Deviation	Very Satisfied (1)	Satisfied (2)	Neutral (3)	Dissatisfied (4)	Very Dissatisfied (5)
Ease in data access	3.21	1.09	5 (3%)	56 (35%)	13 (8%)	77 (47%)	12 (7%)
Data quality	3.19	1.07	11 (7%)	42 (26%)	22 (13%)	82 (50%)	6 (4%)
Productivity improvement	3.44	1.04	4 (2%)	40 (25%)	16 (10%)	87 (53%)	16 (10%)
Quality of reports	3.01	1.07	10 (6%)	53 (33%)	35 (21%)	56 (34%)	9 (6%)
Support for decision-making	3.96	0.99	3 (2%)	18 (11%)	10 (6%)	83 (51%)	49 (30%)
Cost of operation	2.97	0.93	3 (1.5%)	60 (37%)	42 (26%)	55 (34%)	3 (1.5%)

Before further analyzing the results, it is worthwhile to find out if any significant differences exist between the type and size of organization and the level of satisfaction of its employees with the current DBMS. A Kruskal-Wallis test was performed to achieve this objective. Kruskal-Wallis is a non-parametric test (distribution-free) used to compare three or more independent groups of sampled data. It uses the ranks of the data rather than their raw values to calculate the statistics (Barnes, 1998). In this test, the hypotheses for the comparison of independent groups are:

H_o: The samples come from identical populations (no significant differences exist)

H_a: They samples come from different populations (significant differences exist)

The significance level set to test the hypothesis was 5%, i.e. if $p<0.05$ it means there is evidence to reject the null hypothesis in favor of the alternative hypothesis. The results of Kruskal-Wallis test are summarized in Table 4.7.

Table 4.7: Summary of Kruskal-Wallis Test Results for Satisfaction Criteria

Satisfaction Criteria	Organization Type		Organization Size	
	Sum of Squares	p	Sum of Squares	p
Ease in data access	18.69	0.01	7.09	0.13
Data quality	7.06	0.13	6.96	0.12
Productivity improvement	5.49	0.17	6.06	0.11
Quality of reports	11.79	0.01	8.14	0.14
Support for decision-making	4.49	0.19	5.54	0.09
Cost of operation	7.19	0.12	6.36	0.10

The results shown in Tables 4.6 and 4.7 are collectively discussed below with reference to satisfaction criteria:

Ease in data access: Thirty-seven percent respondents were either very satisfied or satisfied with their current DBMS while 54% respondents indicated their dissatisfaction. The results are independent of the organization size however they do vary significantly with the organization type. It was further found that most respondents from departments of transportation and port authorities were satisfied with their current DBMS.

Data quality: Thirty-three percent respondents were found to be satisfied with the data quality while 54% were not satisfied. The results do not vary significantly with the organization size and type which indicates that the data quality issues are present in all types and sizes of organizations and this issue should be given due consideration.

Productivity improvement: Among the surveyed group, 63% respondents had the opinion that their current DBMS is not helping them to enhance their job productivity. The results are again independent of the organization type and size.

Quality of reports: The results show a tie with 39% respondents on either side while 20% stayed neutral. The results seem to vary with the organization type but still independent of the organization size. Further analysis of the data reveals that majority of satisfied respondents were from those organizations who have implemented a sophisticated database management system such as Oracle™, SAP™ or even MS Access™. These organizations include departments of transportation, port authorities, oil and gas companies and transit agencies.

Support for decision-making: The majority of respondents (81%) had the opinion that their current DBMS is not capable to provide sufficient support for decision-making. Like productivity improvement issue, the results are independent of the organization type and size.

Cost of operation: The respondents were asked to indicate their level of satisfaction with the costs involved in managing and operating the current DBMS. The results indicate a tie with 39% respondents on either side while 26% were undecided. The results do not vary significantly with the organization type and size.

In nutshell, it is found that the majority of respondents were not satisfied with their current database management system due to three main reasons, (1) no support for decision-making; (2) no assistance in productivity improvement, and (3) poor data quality. These three issues are considered in detail in the following questions.

4.3.2.4 Decision-Support in Existing Information Systems (IS)

To explore the effectiveness of the existing information systems (IS) for planning and decision-support, the respondents were asked to provide feedback about four predefined functionalities which are very often required by planners and decision-makers. Table 4.8 illustrates the brief results.

Table 4.8: Planning and Decision-Support Functionalities of the Existing IS

IS Functionality	Available in Current System	Not Available
Data integration	39 (24%)	124 (76%)
Generate data trends	32 (20%)	131 (80%)
Generate data summaries	90 (55%)	73 (45%)
Perform "What-If"analysis	6 (4%)	157 (96%)

The Kruskal-Wallis test was performed to determine any significant differences between the organization type, size and the above-mentioned functionalities of the existing information systems. The results are summarized in Table 4.9.

Table 4.9: Summary of Kruskal-Wallis Test Results for Functionality Criteria

Functionality Criteria	Organization Type		Organization Size	
	Sum of Squares	p	Sum of Squares	p
Data integration	5.96	0.16	6.39	0.12
Generate data trends	6.36	0.13	6.96	0.13
Generate data Summaries	7.56	0.11	4.26	0.17
Perform "What-If"analysis	4.39	0.15	5.34	0.15

It is clear from the above analysis that the majority of respondents felt that their existing information systems (IS) do not provide the functionalities often required in planning and decision-making operations expect generating data summaries. The results are independent of the organization type and size. It further reflects that the existing information systems (IS) employed in most organizations are transaction-oriented and not capable of providing any analytical capabilities to the users.

4.3.2.5 Level of Utilization of Existing IS in Planning and Decision-Making

The purpose of this question was to explore the current level of utilization of information systems (IS) in planning and decision-making. The respondents were asked to indicate their level of agreement or disagreement with the given statements. The results of this question are shown in Table 4.10.

Table 4.10: Level of Utilization of Existing IS in Planning and Decision-Making

Statement	Mean	Standard Deviation	Strongly Agreed (1)	Agreed (2)	Neutral (3)	Disagreed (4)	Strongly Disagreed (5)
Use of existing IS to making everyday decisions	3.18	1.15	10 (6%)	53 (33%)	9 (5%)	80 (49%)	11 (7%)
Use of IS for short and long term planning	3.40	1.01	2 (1%)	45 (28%)	14 (9%)	90 (55%)	12 (7%)
Use of IS to prepare summarized reports	2.51	1.12	27 (17%)	70 (49%)	6 (4%)	58 (29%)	2 (1%)

The results of Kruskal-Wallis test for organization type and size are shown in Table 4.11.

Table 4.11: Summary of Kruskal-Wallis Test Results for IS Utilization

Level of Utilization of Existing IS	Organization Type		Organization Size	
	Sum of Squares	p	Sum of Squares	p
For everyday decisions	17.83	0.01	18.99	0.02
For short and long term planning	22.15	0.00	11.06	0.02
To prepare summarized reports	11.63	0.02	16.12	0.02

The following conclusions can be drawn from the above-mentioned results:

Use of existing IS to make everyday decisions: The results indicated that 39% organizations utilize information systems (IS) to make everyday decisions while 56% organizations do not take this advantage. The reason might be that their information systems (IS) do not provide required functionalities as discussed in the previous section. The results vary considerably with the type and size of organization. The large-size public agencies such as departments of transportation, port authorities, transit agencies etc. were found to notably utilize existing information systems (IS) to make most decisions while small-size organizations such as public works departments and many private owners were found to rely more on experience and get-feelings.

Use of IS for short and long term planning: Twenty-nine percent (29%) organizations regularly utilize information systems (IS) for short and long term planning while 62% organization indicated no such utilization. Again, the results are significantly different with respect to organization type and size. The reasons mentioned in the earlier paragraph strongly apply here as well.

Use of IS to prepare summarized reports: As expected, 66% organizations were effectively utilizing information systems (IS) to prepare summarized reports while 30% were not taking advantage of this functionality of information systems (IS). Apparently the later were small-size organizations who are managing most part of their data manually. This claim is further supported by the results of Kruskal-Wallis Test which shows a significant difference with respect to organization type and size.

4.3.2.6 How to Improve Existing Information Systems (IS)?

At the end of section 2, the respondents were asked that if they would like to improve their existing information systems (IS), which solution they would prefer. They were given the choice to select one option among the four given options. The results are depicted in Table 4.12.

Table 4.12: How to Improve Existing Information Systems?

Available Option	Response Rate
Hiring additional staff to more effectively utilize the existing IS	15 (9%)
Implementing a new and enhanced IS	56 (34%)
Both hiring additional staff and implementing new IS	86 (53%)
Other (training, outsourcing, integrating systems etc.)	6 (4%)

More than half of the respondents (53%) indicated that they would like to implement a new information system (IS) and would also hire additional staff to operate it. The second highest group of respondents (34%) had the opinion to implement a new and enhanced information system (IS) only. The breakdown clearly indicates that most of the organizations are not completely satisfied with their existing information systems (IS) and have the opinion to replace it with a better and more enhanced system.

4.3.3 Data Warehousing Implementation in the Respondent's Organization

After presenting a brief overview of the data warehousing technology, its advantages and usefulness in planning and decision-support, the respondents were asked if data warehousing is needed in their organization. Their responses are shown in Figure 4.8.

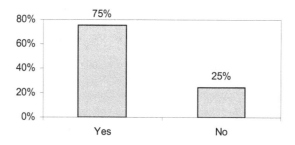

Figure 4.8: Respondents Opinion about Data Warehousing Implementation

Out of 163, 123 (75%) respondents had the opinion that data warehousing is needed in their organization while 40 (25%) respondents felt that their organization do not need this technology. Kruskal-Wallis test results (X^2 = 17.34, $p<0.01$) indicated that the later organizations are small size departments with 100 or less employees. No significant difference was found with respect to organization type (X^2 = 4.95, $p<0.13$).

The respondents who considered data warehousing not suitable for their organization were further asked about the main reasons behind their opinion. The reasons provided by them are shown below in the descending order.

(1) Size of organization is small (38 respondents)

(2) Trend analysis is not important (31 respondents)

(3) Cost is high as compared to its benefits (27 respondents)

(4) Executive management is not interested to invest in new IS (8 respondents)

(5) Implementation time is long (4 respondents)

(6) Data ownership problems (2 respondents)

These reasons indicate that data warehousing is more appropriate for medium to large size organizations that deal with large volumes of data and are willing to invest in new information technology solutions.

At the end, the respondents' feedback was collected about the evaluation practices used in their organizations to decide about investment in new information technology solutions. The responses are shown below in the priority order.

(1) By comparing competitive advantages with the existing softwares or IS (76 respondents, 49%)

(2) Economical analysis such as Cost/benefit, Return on Investment, Value Analysis etc. (42 respondents, 27%)

(3) Both techniques (43 respondents, 28%)

(4) Other methods (2 respondents, 1%)

This response indicates that half of the organizations (i.e. 50%) do not perform any formal economical analysis to calculate the value and worth of applied IT solution to their organization. Anecdotal evidence would suggest that this finding is not due to a lack of knowledge of the available techniques instead it is apparent that IT does not form an integral part of their business strategy for competitive advantage. Before investing in costly IT solutions like data warehousing, organizations must perform some sort of economical analysis to calculate long term benefits. The data warehousing is a costly

solution and its economic benefits start to appear after few years of implementation.

4.4 Summary

An assessment of the data management practices employed in construction owner

organizations and degree of utilization of existing information systems (IS) in planning

and decision-making is presented in this chapter. It is found that medium to large size

construction owner organizations use a formal database management system to store and

manage project data while small size organizations rely on the traditional paper and

folder techniques. Whether managing data manually or through computerized means,

most organizations are not taking full advantage from their data. The main utilization of

project data was found in preparing summarized reports for project monitoring and

control. Hardly few organizations are utilizing their project data in formal planning and

decision-making operations. Most organizations indicated their willingness to adopt data

warehousing as a promising solution to address their current data management and

decision-support needs. The analysis presented in this chapter also validated the initial

research hypothesis that construction owner organizations do not effectively utilize

project data for planning and decision-making due to lack of decision-support in their

existing information systems (IS).

CHAPTER 5

MODELING OF CONSTRUCTION OWNER ORGANIZATIONS

5.1 Introduction

The purpose of owner organization modeling is to prepare a series of discrete models that could capture the existing or "AS-IS" state of the organization from different perspectives such as functional, organizational, informational, decision-making, etc. These discrete but interrelated models can be used to understand the organization's complex business operations, flow of information within, into and outside the organization, identification of decision nodes and their hierarchy, and data requirements for these decisions. This information is used in this research to solicit user requirements for the development of data warehousing functional model. As stated earlier, the Miami-Dade Transit (MDT) was chosen as a client organization for this modeling. However, the models developed are generic for construction operations and can be used for other owner organizations with little modifications. Figure 5.1 depicts the framework used for the modeling of MDT. First, a brief macro-level study was conducted to understand MDT's nature, work style, culture and business operations. It was followed by assessment of data management practices and degree of utilization of information systems (IS) in planning and decision-making. Next, the organizational, functional, informational and decision models were developed using the CIMOSA framework. In the following step, data and information flow models were prepared to depict the flow of information within, into and outside the organization. At the end, decision-support requirements matrices for different management levels were prepared by mapping data and information flow models with the decision-perspective model.

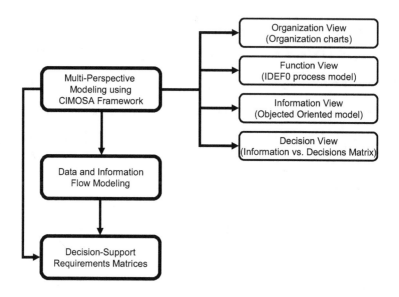

Figure 5.1: Framework Adopted for Construction Owner Organizations Modeling

5.2 An Overview of Miami-Dade Transit

Miami-Dade Transit (MDT) is the 16th largest public transit system in the nation, and the largest transit agency in the state of Florida (Miami-Dade county, 2005). It is responsible for the planning, development and operations of all public transit services in the Miami-Dade county. It operates three transit modes: bus, rail and monorail. The agency carried over 83,619,000 passenger trips in 2004 (Miami-Dade county, 2005).

The organizational structure of MDT is illustrated in Figure 5.2. MDT is a functional organization with three major units named as Operations, Planning and Development, and Administration. The Operations unit is responsible for operations of rail and bus services, field maintenance and customer service. Planning and development unit is in charge of planning and construction of transit-related projects such as road works, bridges, bus and train stations, etc.; installation of equipment and quality control.

84

Administration unit is liable for new business development, audit and financial services and all other administration-related internal and external affairs. Each of these three units has several divisions responsible for more specialized operations. MDT is headed by an Executive director appointed by the Miami-Dade county. Every unit has a Deputy Director who supervises division chiefs. Division chiefs are supported by managers, section managers and various office and field staff.

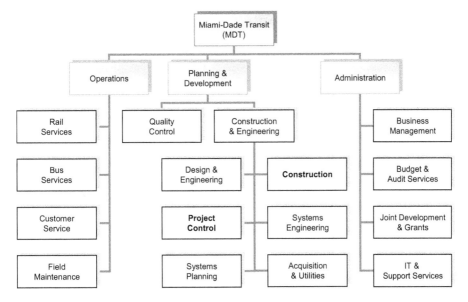

Figure 5.2: Organizational Structure of Miami-Dade Transit (MDT)

The modeling efforts in this research were focused on the Project Control division and the Construction division because these two divisions are directly involved in the planning and development of all construction-related projects. The Design and Engineering division, Systems planning division and the Quality control division coordinate with these two divisions as well as other divisions during the project planning, execution and control stages. Hence these divisions were partially considered in the

85

model development efforts. Table 5.1 provides a brief overview about the nature of construction projects undertaken by the MDT.

Table 5.1: An Overview of Construction Projects undertaken by the MDT

Types of undertaken construction projects:	Infrastructure development (roads, bridges, transit facilities)
Number of projects undertaken during the last 5 years:	Approximately 30
Annual construction expenditures:	$75 - $100 Million

5.3 Data Management Practices at MDT

To evaluate the data management practices at MDT and degree of utilization of information system in planning and decision-making, a questionnaire survey similar to the ones mentioned in Chapter 4 was conducted. The purpose of conducting this survey was to investigate the data management practices at micro-organization level. The target population was strategic, control and functional management in the Planning and Development unit. The questionnaire design was similar to the ones shown in Appendix A. The only difference was the exclusion of section 1 as the organization characteristics were already known. The questionnaires were handed over to 12 personnel and responded by all thus yielding a response rate of 100%. The results of section 2 and 3 of the questionnaire survey are discussed in the following paragraphs.

The data management and storage practice was found to be 50% manual and 50% computerized. The main softwares used for these operations include MS Excel™, MS Access™, and Primavera™. Although Primavera™ is not a data management software but it has a database component. It was found that the scheduling and cost data were archived in Primavera™ and important data were exported to MS Excel™ for further in-depth analysis.

Table 5.2 shows the level of satisfaction of executives with the current database management system. Due to small size of the sample, respondent's response was evaluated in numbers instead of percentage. The results indicate that the majority of executives had opinions not in favor of their current database management system (DBMS). A comparison of these results with the ones obtained from other organizations, as shown in Table 4.6, indicated that dissatisfaction of executives with their current data management practices at most owner organizations lied at the same level.

Table 5.2: Level of Satisfaction with the Current Database Management System

Satisfaction Criteria	Mean	Standard Deviation	Very Satisfied (1)	Satisfied (2)	Neutral (3)	Dissatisfied (4)	Very Dissatisfied (5)
Ease in data access	3.58	1.09	--	3	1	6	2
Data quality	3.08	1.06	1	2	2	7	--
Productivity improvement	3.83	0.93	--	1	3	5	3
Quality of reports	3.16	0.93	--	4	2	6	--
Support for decision-making	4.25	0.87	--	1	--	6	5
Cost of operation	2.58	1.31	3	4	--	5	--

Table 5.3 summarizes the results about planning and decision-support functionalities of existing information systems (IS) at MDT while Table 5.4 delineates findings about level of utilization of existing information systems (IS) in planning and decision-making.

Table 5.3: Planning and Decision-Making Functionalities of the Existing IS

DBMS Functionality	Available in Current System	Not Available
Data integration	9	3
Generate data trends	10	2
Generate data summaries	7	5
Perform "What-If" analysis	12	--

Table 5.4: Level of Utilization of Existing IS in Planning and Decision-Making

Statement	Mean	Standard Deviation	Strongly Agreed (1)	Agreed (2)	Neutral (3)	Disagreed (4)	Strongly Disagreed (5)
Use of existing IS to make everyday decisions	3.25	1.14	1	3	--	8	--
Use of existing IS for short and long term planning	3.83	1.03	--	2	1	6	3
Use of existing IS to prepare summarized reports	2.08	1.24	5	4	--	3	--

It is recognized from the results shown in Tables 5.3 and 5.4 that at MDT, information systems (IS) were not effectively utilized in planning and decision-making operations. The main reason was these information systems (IS) did not store and organize data in forms and formats which were required by the decision makers.

When asked about how improvements could be made in the existing information systems (IS), 10% respondents supported the choice of hiring additional staff, 22% backed the option of purchasing a new information system while 68% were in favor of both hiring additional staff as well as purchasing a new information system.

In terms of data warehousing implementation at MDT, 10 out of 12 respondents voted in favor of this option. The 2 opposition votes showed their disinterest due to long implementation time and data ownership problems. All the respondents indicated that no formal IT product evaluation method was used within the organization and usually decisions were made by comparing the competitive advantages of the new system with the existing ones.

The results of this questionnaire survey revealed that MDT is one of the typical owner organizations where project data are not effectively utilized for planning and decision-making. Hence MDT can be taken as a good candidate organization for testing and implementation of the proposed decision-support framework.

5.4 Multi-Perspective Modeling Using CIMOSA Framework

The multi-perspective modeling was performed using the Computer Integrated Manufacturing Open Systems Architecture (CIMOSA) framework. As mentioned in section 3.4.2, three modeling perspectives (i.e. Organization, Function and Information) of the original CIMOSA framework were selected while a forth new perspective of Decision was introduced to model decision-making process in the construction owner organizations. The CIMOSA framework proposes organization modeling for three phases namely Requirements Definition, Design Specification and Implementation Description (Vernadat, 1993). In this research, our main purpose was to understand the existing state of the organization and capture user requirements; hence only first phase named Requirements Definition was conducted. Figure 5.3 illustrates the modified CIMOSA framework adopted in this research.

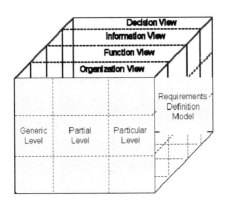

Figure 5.3: Modified CIMOSA Framework

Before modeling the organization from four above-mentioned perspectives, a domain level diagram of MDT's construction operations system was prepared. A domain defines a subset of the organization performing a specific task and is particularly useful in breaking down a system complexity (Vernadat, 1993). The purpose of this diagram was to identify the construction-related functional areas at MDT and delimit them from the other organization's operations. Figure 5.4 shows the domain level diagram of MDT's construction operations indicating various domain processes and flow of information between these domains.

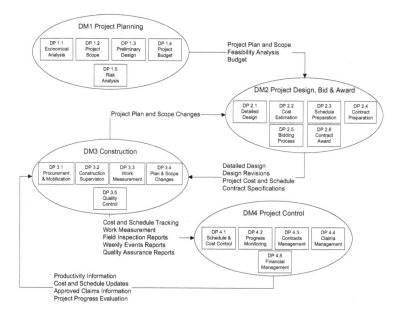

Figure 5.4: Domain Level Diagram of MDT's Construction Operations System

As shown in Figure 5.4, the MDT's construction operations system can be broken down into four discrete domains of Project Planning, Project Design, Bid & Award, and Construction and Project Control. Each domain has various domain processes which represent organization's main business processes. The arrows indicate flow of information between these domains. This diagram helped to identify the different construction-related operations (domain processes) performed in the organization and served as a guide map in the preparation of CIMOSA-based models.

5.4.1 Organization View

The organization view helps to understand the organizational structure and interdependence of various organizational units. The organization charts are generally used to describe the organization view. These charts can be broken down at different levels to explain the macro and micro level details of various functional and operational units. For MDT's modeling, organization charts were prepared at two levels of detail. The first chart which is already shown in Figure 5.2 illustrates the major units and divisions at a macro level. Since two divisions namely Project Control division and the Construction division are mainly involved in construction-related operations, the micro level organization details of these two divisions are depicted in Figure 5.5.

Each division is headed by a division chief who is responsible for all major implantation and control decisions within that division. The division chief is supported by Managers (level-3) who are in charge of different functional units and make project-level functional decisions in consultation with the division chief. The managers are the main coordination agents with different stakeholders such as contractors, consultants, vendors as well as other division's units. The managers supervise section managers (or level-2 managers) who are responsible for executing and supervising various construction activities in the design office as well as in the field.

91

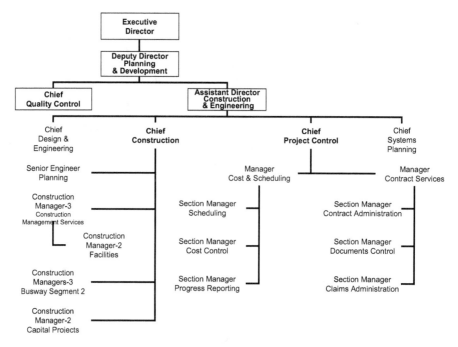

Figure 5.5: Organization Structure of MDT's Construction Related Divisions

5.4.2 Function View

The function view shows organization processes and behavior encapsulated in one model. The function represents what process elements (activities) are being performed while the behavior depicts when activities are performed as well as aspects of how they are performed through feedback loops, decision-making conditions, and entry and exit criteria (Giaglis, 2001).

A number of techniques are available for modeling business functions and behavior such as Flowcharting, IDEF0 technique, Petri nets, Simulation, Role activity diagramming, etc. For this research, the IDEF0 (Integrated Definition for Functional Modeling) technique was selected due to the following reasons: (1) it incorporates decisions, actions and activities of an organization in the functional model; (2) it is simple and uses only one notational construct, the ICOM (input-control-output-mechanism); and (3) it supports functional modeling by progressively decomposing higher-level models into more-detailed models that depict the hierarchical decomposition of activities (Giaglis, 2001).

The basic notations used in IDEF0 modeling are shown in Figure 5.6. A process is shown in a process model as a rectangular box. It contains a unique text description or label that describes what the process is. The label is expressed as a verbal phrase because a process is an action. Arrows are constraints (input, output, control and mechanism) that define the box. The input data (on the left) are transformed into output data (on the right). Controls (on the top) govern the way an activity is performed. Mechanisms (on the bottom) indicate the means by which a function is performed. IDEF0 models are defined to represent the process at various levels of abstraction. Higher level models are usually very abstract and can be decomposed into levels of increasingly greater detail. The level of detail is determined by the modeling requirements (Chen et al., 2004). This concept of hierarchical decomposition is shown in Figure 5.7.

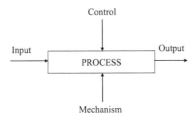

Figure 5.6: IDEF0 Modeling Notations

93

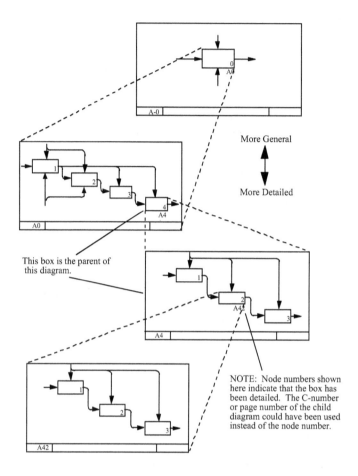

Figure 5.7: Basic Concepts of IDEF0 Modeling (Adapted from Giachetti, 2002)

After analyzing the MDT's domain level diagram (Figure 5.4), it was decided to decompose the Functional model up to the 3rd hierarchical level thereby modeling domain processes, enterprise activities and enterprise tasks. This level of decomposition was considered as sufficient by other researchers to model the organization's functional processes (Chen et al., 2004, Lykins et al., 1998). Figure 5.8 illustrates the node tree showing different decomposition levels used for MDT's functional modeling.

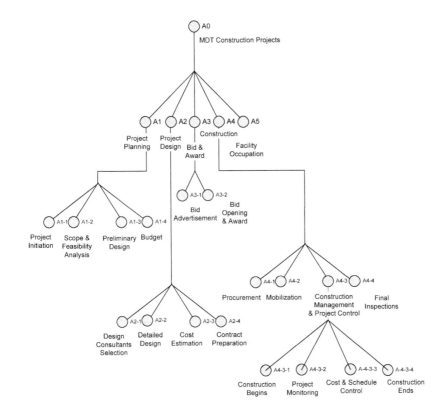

Figure 5.8: Node Tree showing Decomposition Levels used for MDT's IDEF0 Modeling

IDEF0 models prepared for each decomposition level are shown in Figures 5.9-5.15. For preparing IDEF0 models, each division of MDT was thoroughly studied; interviews were conducted with the division managers to investigate the work flow and work scope of each division. Different project documents such as proposals, work orders, contracts documents and project reports were also examined. The prepared models were shown to the division managers and their feedback was collected for necessary refinements.

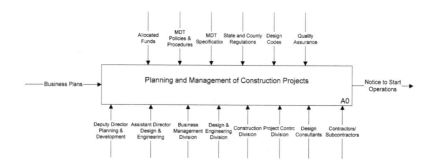

Figure 5.9: IDEF0 Model of MDT at Level 0 (Maximum Abstraction)

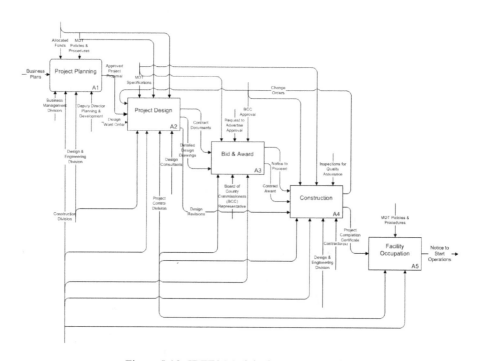

Figure 5.10: IDEF0 Model of MDT at Level 1

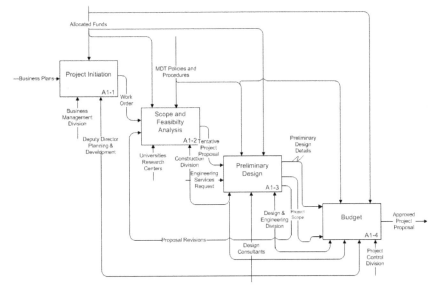

Figure 5.11: IDEF0 Model of Project Planning Phase (Level 2)

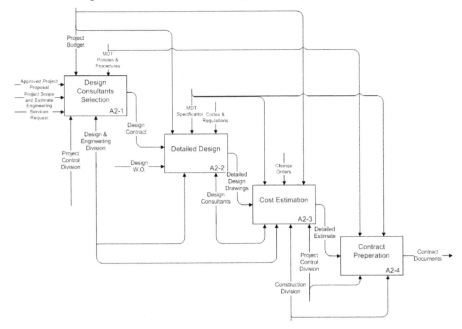

Figure 5.12: IDEF0 Model of Project Design Phase (Level 2)

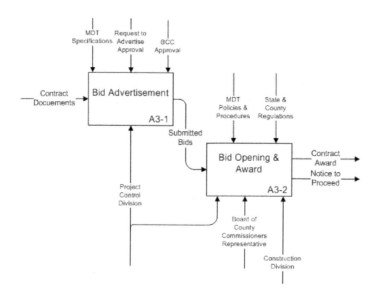

Figure 5.13: IDEF0 Model of Bid & Award Phase (Level 2)

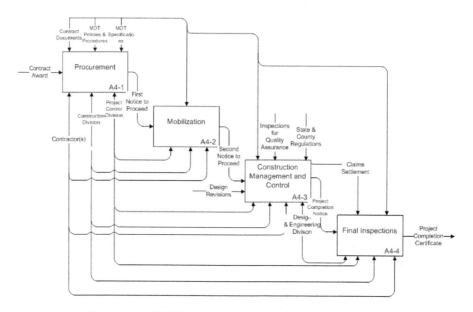

Figure 5.14: IDEF0 Model of Construction Phase (Level 2)

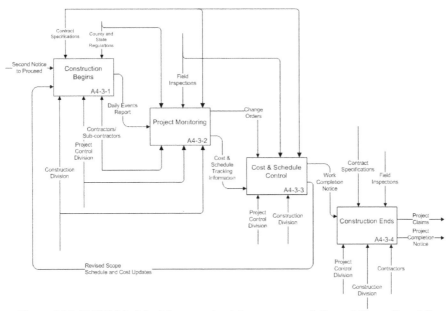

Figure 5.15: IDEF0 Model of Construction Management and Control Phase (Level 3)

The following information was revealed from these models about the work flow at MDT. A new project is initiated by the Business Management division in consultation with the Deputy Director, Planning and Development (P&D). The Deputy Director P&D issues a work order for scope and feasibility analysis and send the initial proposal to the Planning and the Construction divisions. These divisions evaluate the economic feasibility of the project and determine its scope. If the project is feasible then the proposal is sent to the Design & Engineering division which prepares the preliminary project design and sends the design to the Project Control division for budget preparation. The Project Control division prepares cost estimates, project schedule and related contract documents and advertise the project for bidding. The bidding is carried out in the presence of a Board of County Commissioners Representative to ensure transparency. All the contractors who take part in the bidding should be prequalified MDT contractors. The award decision is based on the submitted value of the bid, project difficulty level and past performance of the contractor in MDT's or other similar projects. After the project award, the contractor is asked to submit its own schedule of activities with greater level of detail. Once the construction is started, the Construction division continuously monitors the construction operations, prepare daily and weekly field reports and send them to the Project Control division. The Project Control division (PCD) is solely responsible for Project Management and Control functions such as schedule control, cost control, measuring project progress, project productivity and claims administration. The Quality Control division monitors the construction quality in consultation with the Construction division. Once the project is completed, final inspections are carried out by the Construction division and if satisfied, a Project Completion Certificate is issued and the facility could be occupied for transit operations.

The IDEF0 models indicated that most of the construction-related data are collected and analyzed by the Construction division and Project Control division; hence the further modeling is particularly focused on these divisions.

5.4.3 Information View

The information view represents the information entities (data) produced or manipulated by a process and their interrelationships (Giaglis, 2001). It describes what data are produced during a process and how the data are stored by an organization in one or many information systems. Oliver et al. (1997) mentioned that a function model together with an information model, define what is the "right" information for designing a new information system. This means that information models are needed, in addition to function models, to provide a comprehensive view of information and to achieve effective and efficient communication. Besides, a good information model minimizes redundancy by capturing common characteristics of a system in a single place. This capability improves comprehension and reduces the risk of error (Lykins et al., 1998).

A number of techniques are available for modeling information view such as Data flow diagramming, Entity-relationship diagramming, State-transition diagramming, IDEF techniques (IDEF1x) and Unified Modeling Language (Giaglis 2001). After analyzing the suitability of different techniques with the objectives of this research, it was decided to use Unified Modeling Language (UML) to capture information view. The main reason was that UML models information from an object-oriented perspective. In the IDEF0 function model, the information objects flow in the pipe-lines of the arcs and can be best represented using UML Class diagrams. A UML Class diagram depicts the classes (or categories) within an information model. Each class has attributes (member variables), operations (member functions) and relationships with other classes (Booch et al., 1999). Figure 5.16 represents the UML Class diagram syntax and different types of relationships which may exist between the classes.

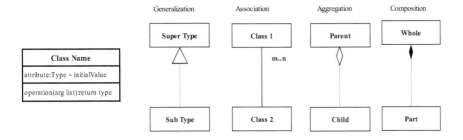

Figure 5.16: (Left) Syntax of a Class Diagram, (Right) Relationships between Classes

For preparing UML Class diagram for MDT, different processes from domain level diagram (Figure 5.4) and DIEF0 functional models (Figures 5.9-5.15) were identified. The data generated through these processes were categorized to define data classes. It was decided to prepare a single comprehensive Class diagram for the entire system instead of generating separate diagrams for different domains because most of the data are shared between these domains. The diagram is independent of division or department boundaries and focused only on the data which are generated during different construction-related operations. Such a single Class diagram helped to reduce the data redundancies and gave a holistic view of the entire organization's data. Figure 5.17 shows the UML Class diagram prepared to represent information view of MDT's construction operations system.

This diagram categorizes the construction data into six major categories namely Project Information, Project Plan, Contracts Information, Project Control Information, Financial Information, and Safety and Quality Information. These categories indicated the main subjective areas for data management and later used as a key for locating the subjects for the data warehousing functional model.

It was also found that there was no standard project management information system at MDT and different divisions stored same data in different formats and in different applications. The data flow between these divisions were mostly "on request" except the periodic flow of progress reports. All the data flows were asynchronous in nature. No IT infrastructure was found for electronic data sharing within or outside MDT.

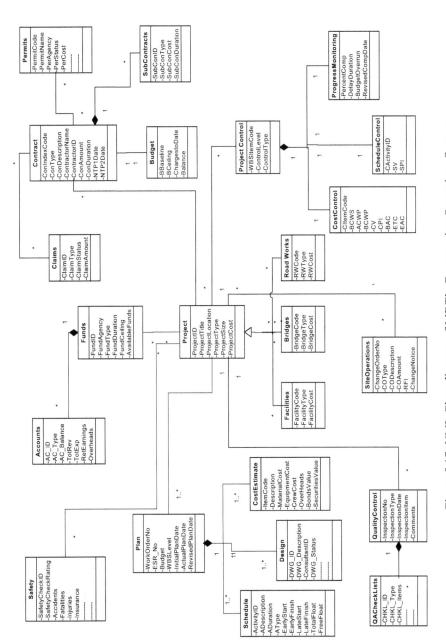

Figure 5.17: UML Class diagram of MDT's Construction Operations System

103

5.4.4 Decision View

The decision view represents the decision-making process in the organization at different management levels and identifies the information needs for each type of decision. It explicitly identifies "who makes what type of decisions" and "what are the data or information requirements for these decisions" (Robbins, 2003).

For modeling the decision view, first of all it is important to identify the different management levels within the organization. As suggested by Tang et al. (2004), a four level management hierarchy was used which divided the management into Strategic, Implementation or Control, Functional or Operational and Transactional levels. Figure 5.18 identifies these management levels at MDT. The information presented about the percentage of decisions made at each management level is approximate and based on the subjective judgment.

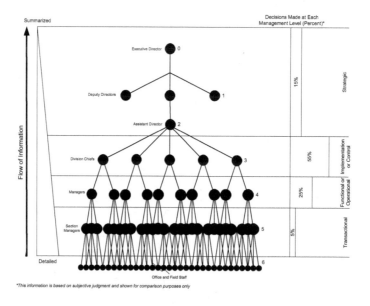

Figure 5.18: Management Levels for Decision-Making at MDT

Figure 5.18 also depicts that approximately 70% decisions are made at strategic and control management levels, hence it is imperative to focus decision-support framework development efforts on these two management levels. Besides, the executives at these management levels need more summarized information and trend analysis which are currently not available in the existing information systems (IS) at MDT as explained in section 5.3.

Different techniques are available for modeling the decision perspective such as Decision trees, Analytical hierarchy process, Criteria and decision matrices, Cause and effect diagrams, etc. (Robbins, 2003). None of these methods explicitly satisfies the research needs because our focus was on identifying decisions with their corresponding information requirements. Hence it was decided to use the concept of Criteria and Decision matrix and modify it to develop a new "Construction decisions v. Information needs" matrix. For developing this matrix, interviews were conducted at different management levels to identify construction-related decisions and information requirements for these decisions. The developed matrix is presented in Figure 5.19. Since most of these decisions are typical in construction organizations, hence this matrix could be used for other construction owner organizations with minor modifications.

In the developed matrix, the black dots represent the information requirements for each type of decision. These information requirements represent an ideal case scenario; however in reality all information is not always available. Tables 5.6-5.10 presented at the end of this chapter shows which type of information is usually available to the executives and which information is provided on request after analyzing the necessary data.

Decision Level	Management Level	Construction Decisions / Information Needs	Economical & Financial			Planning					Monitoring and Control							Detailed and Summarized Reports			
			Economic Analysis (NPV, ROI, IRR, Dis. Cash Flow)	Financial Analysis and Projections	Miami-Dade County and MDT Budget	MDT Specifications, Policies and Procedures	Cost Estimates (Screening, Semi-detailed and Detailed)	Precedence Analysis	Risk and Uncertainty Analysis	Contractors Performance History	Contract Specifications, Design Codes	Productivity Analysis	Variance Analysis	Sensitivity Analysis	S-Curve Analysis	Work Measurement	Legal Information	Site Inspections	Field Reports, Photographs and Videos	Project Progress	Information from Past Projects Archive (e.g. Data Trends)
Strategic	Deputy Directors Assistant Director	New Projects and Services			●																●
		Funds Allocations		●	●																
		Annual Projects Progress Evaluation		●																●	
Implementation or Control	Division Chiefs	Feasibility and Scope	●			●			●					●							
		Project Final Plan				●	●	●			●										
		Project Specifications				●					●						●				●
		Resources Allocation					●	●													●
		Bidding Method and Procedures				●				●							●				
		Contract Award								●											●
		Financial Management		●	●							●	●		●						
		Risk and Uncertainty Issues							●					●						●	
		Project Plan and Scope Changes	●	●							●		●	●				●		●	
		Quality Control																●	●		●
		Contractors Performance										●	●		●	●		●	●		
		Claims and Negotiations				●					●					●	●	●	●		●
		Monthly Projects Progress Evaluation		●								●	●		●	●				●	
Functional or Operational	Managers	Project Design Selection				●					●										●
		Construction Methods Selection				●					●										●
		Cost and Time Allocations					●	●			●										
		Cost and Schedule Control											●		●						
		Contracts Monitoring				●					●							●	●	●	
		Procurement Management		●		●	●														●
		Change Orders Approval		●							●		●		●				●		
		Weekly Projects Progress Evaluation		●								●	●		●	●			●	●	
Transactional	Section Manages	Specifications and Compliance Issues				●					●							●			
		Site Operations and Problems				●					●										●
		Daily Project Progress Evaluation														●					

Figure 5.19: Construction Decisions v. Information Needs Matrix

5.5 Data and Information Flow Modeling

The purpose of data and information flow modeling was to investigate their flow

patterns in the organization. It helped to identify operational data and information flow

between different divisions within MDT, and between MDT and external stakeholders

such as contractors, consultants and vendors.

The information flow modeling was performed at two levels of detail. The top-level

model as shown in Figure 5.20 considered the entire construction operations system at

MDT. It represents information flow between different MDT divisions and external

stakeholders. The bottom-level model was developed for individual divisions.

Construction and Project Control divisions were modeled at this level as shown in

Figures 5.21 and 5.22. The Design & Engineering division was not involved in major

construction operations and hence not considered at this stage. In these information flow

models, the different divisions/stakeholders are represented by rectangles while the flow

of information is symbolized by dotted arrows.

The following information is conceived from these models. The section managers

receive operational data from the field staff of MDT, contractors and consultants. They

analyze data and send the analysis results to the managers. The managers integrate

information received from different section managers and review the project state. Based

on their decision authority, they either take the decision or send the request to the division

chief for making appropriate decisions. The managers are also responsible for preparing

different project reports at monthly basis and sending them to the division chiefs. The

division chiefs evaluate the project performance on monthly basis. They integrate the

information on quarterly and yearly basis and send it to the strategic management.

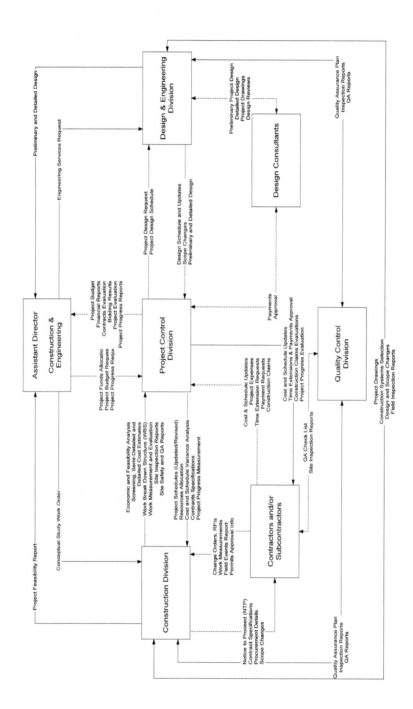

Figure 5.20: Information Flow Model of Construction Operations System at MDT

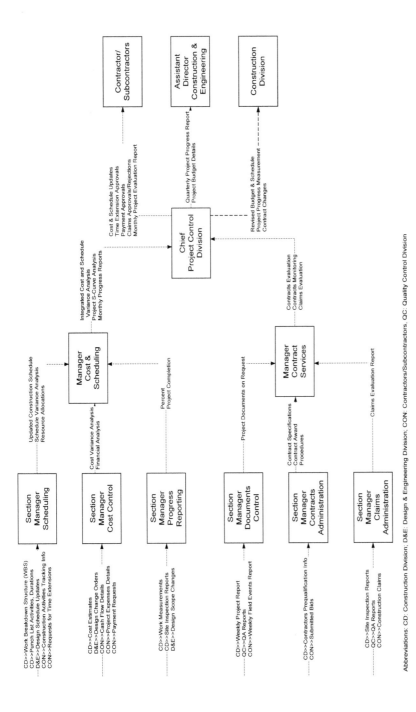

Figure 5.21: Information Flow Model for Project Control Division

Abbreviations: CD: Construction Division; D&E: Design & Engineering Division; CON: Contractors/Subcontractors; QC: Quality Control Division

109

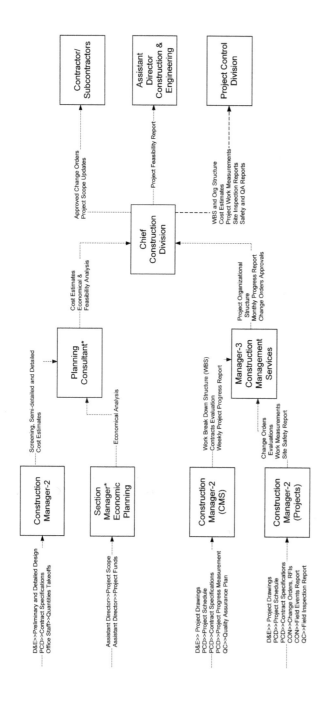

Figure 5.22: Information Flow Model for Construction Division

Abbreviations: PCD: Project Control Division; D&E: Design & Engineering Division; CON: Contractor/Subcontractors, QC: Quality Control Department
* not a part of the construction division

110

Table 5.5 illustrates the classification of project reports prepared for different management levels. It indicates that strategic and control management who are responsible for making approximately 70% project decisions do not have direct access to detailed and exceptional project reports. At many occasions, the executives might be interested in drilling-down and drilling-across the data to quickly find specific information. Their existing reporting system does not provide this functionality and as a result they need to request for this information from managers and section mangers. This information gathering process takes time and could have negative impact in critical project situations.

Table 5.5: Classification of Project Reports Prepared for Different Management Levels

No.	Report Title	Preparing Division	Annual	Bi-annual	Quarterly	Monthly	Weekly	Daily	Format
			Distribution Frequency						
1.	Projects Progress	Project Control Construction	✓	✓	✓	✓	✓	✓	Text, Spreadsheet
2.	Budget	Project Control	✓	✓	✓	✓			Spreadsheet
3.	Finance	Project Control	✓	✓	✓	✓			Spreadsheet
4.	Schedule Control	Project Control			✓	✓			Spreadsheet
5.	Cost Control	Project Control			✓	✓			Spreadsheet
6.	Project Events	Construction				✓	✓	✓	Spreadsheet
7.	Field Inspection	Construction				✓	✓	✓	Text
8.	Quality Control	Quality	✓		✓	✓			Spreadsheet
9.	Design Changes	Design & Engineering				✓	✓	✓	Spreadsheet
	Prepared for Management Level		Strategic			Control	Functional	Transactional	

Besides information flow models, data flow models were also prepared for Project Control and Construction divisions. The purpose of a data flow model is to examine how operational data flows within a division, what processes are used to convert data into required information and where these data and information are stored.

Data flow diagrams (DFD) were used for data flow modeling. A DFD graphically depicts the flow of data among external entities, internal processing steps, and data storage elements in a business process. DFDs are used to document systems by focusing on the flow of data into, around, and outside the system boundaries (Giaglis, 2001). In a DFD, a straight rectangle represents an external agent (departments, stakeholders), a curved rectangle represents process and an open rectangle shows data storage. Figures 5.23 and 5.24 show the DFDs for Project Control and the Construction divisions.

Figure 5.23 illustrates the data flow for project control operations such as schedule control, cost control and project progress monitoring. It indicates that Project Control division receives most of the operational data from Construction division and Contractors. No formal checks are performed to validate the data. Poor quality data could lead to misleading results. This is another reason that executives at MDT do not solely rely on the data and make many decisions based on their experience and gut-feelings. Figure 5.24 shows the DFD for the Construction division. A close look at both figures indicates that data are exchanged between the two divisions at section managers' level. A further investigation revealed that these data exchanges are mostly done through external storage media and via e-mail attachments. When this research was conducted, these divisions did not use any formal methods for data exchange such as Electronic Data Interchange (EDI) or use of Wrappers to interconnect two or multiple applications.

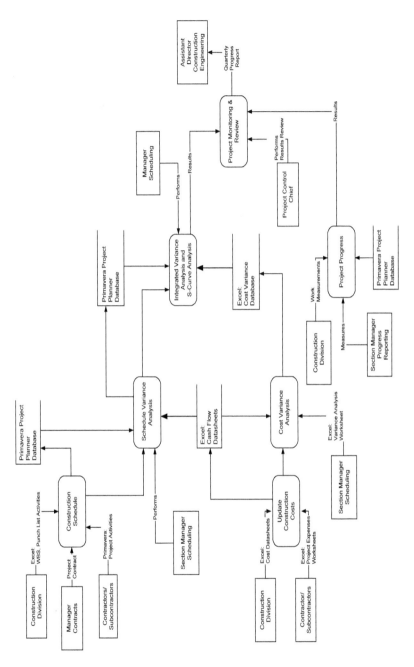

Figure 5.23: Data Flow Diagram for Project Control Division

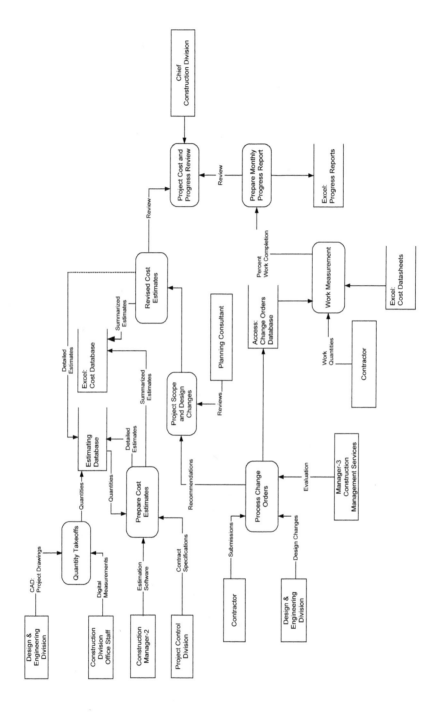

Figure 5.24: Data Flow Diagram for Construction Division

5.6 Development of Decision-Support Requirements Matrices

The Construction decisions versus Information needs matrix (Figure 5.19) and Data and Information flow diagrams (Figures 5.20-5.24) were mapped to prepare decision-support requirements matrices for strategic, control and functional management of the Project Control division and the Construction division. These matrices indicate the information desired for each construction-related decision and the information which was actually available. They served as the guide map for identifying user requirements to develop the functional model of the data warehouse. Tables 5.6-5.10 illustrate these matrices.

For each decision, the decision maker's satisfaction level about the information available versus information desired was measured subjectively. The executives were asked to indicate their satisfaction on a scale of Low, Medium and High. The decision areas where the decision maker's satisfaction level was low or medium were marked as the potential subject areas. These subject areas were focused in the development of the decision-support framework.

In particular, the following subject areas were considered as candidates at this stage: Project progress, Cost management, Schedule management, Change orders management, Productivity analysis, Past projects details, Contractors performance history and Financial analysis. At a later stage, some of these subject areas were integrated to present more consolidated information. For example, subject areas of Projects progress and Change orders management were integrated to form a new subject area of Project Performance. These details are shown in section 6.2 of chapter 6.

115

Table 5.6: Decision-Support Requirements Matrix for Strategic Management

Decision Level	Decision Node	Units	Decisions	Information Needs	Desired		Available		Satisfaction Level (High/Med/Low)
					Granularity Level	Dimensions	Granularity Level	Dimensions	
Strategic	Deputy Director/Assistant Director	Planning and Development/ Construction & Engineering	New projects and services	Economical analysis	Summarized	Projects, Time, Size, Location	Detailed	Projects, Size, Time, Location	High
			Accounts allocations	Financial analysis	Summarized	Projects, Accounts, Contracts, Time	Summarized	Projects, Accounts, Contracts, Time	High
			Annual projects progress evaluation	New project budget	Summarized	Projects, Contracts	Summarized	Projects, Contracts	High
				Recent project progress	Summarized	Project Type, Cost, Location, Time	Summarized	Project type, Cost	Low
				Past project archive	Summarized	Cost, Project Type, Time, Performance	Summarized	Cost	Medium

Table 5.7: Decision-Support Requirements Matrix for Control Management (Project Control Division)

Decision Level	Decision Node	Division	Decisions	Information Needs	Desired		Available		Satisfaction Level (High/Med/Low)
					Granularity Level	Dimensions	Granularity Level	Dimensions	
Control or Implementation	Chief	Project Control	Resources allocation	Project cost and schedule	Summarized	WBS level, Time	Summarized	WBS level, Time	High
				Financial analysis	Detailed, Summarized	Projects, Accounts, Time	Summarized	Projects, Accounts, Time	Medium to High
			Final contract specifications	Project budget	Detailed, Summarized	Contracts, Time	Summarized	Contracts, Time	Medium to High
			Risk and uncertainty issues	Past project archive	Summarized	Delays, Claims, Cost overruns	NA	NA	Low
			Contract award	Contractors performance history	Summarized	Productivity, Quality	NA	NA	Low
				Past project archive	Summarized	Contracts, Claims	Detailed	Contracts, Claims	Medium
			Project plan and scope changes	Productivity analysis	Summarized	Projects, Contractors, Time	NA	NA	Low
			Contractors performance	Cost and schedule variance analysis	Detailed, Summarized	Projects, Contracts, Time	Detailed	Projects	Medium
			Monthly projects evaluation	S-curve analysis	Detailed, Summarized	Projects, Contracts, Time	Detailed	Projects	Medium
			Claims and negotiation issues	Work measurement and percent completion	Summarized	Projects, Contracts, Time	Summarized	Projects, Contracts, Time	High
			Financial Management	Financial analysis	Detailed, Summarized	Projects, Contracts, Time	Summarized	Projects, Contracts, Time	Medium to High

Table 5.8: Decision-Support Requirements Matrix for Control Management (Construction Division)

Decision Level	Decision Node	Division	Decisions	Information Needs	Desired		Available		Satisfaction Level (High/Med/Low)
					Granularity Level	Dimensions	Granularity Level	Dimensions	
Control or Implementation	Chief	Construction	Project feasibility and scope	Economic analysis	Detailed, Summarized	Project size, Location, Time	Detailed	Project size, Location, Time	High
				Financial analysis	Detailed, Summarized	Projects, Accounts, Time	Summarized	Projects, Accounts, Time	Medium to High
				Past project archive	Summarized	Project, Cost, Time	NA	Project	Medium
			Project final plan	Project budget	Detailed, Summarized	Contracts, Time	Summarized	Contracts, Time	High
			Project specifications (construction related)	Project cost and schedule	Summarized	WBS level	Summarized	WBS level	High
				Past project archives	Summarized	Contracts, Time	NA	NA	Low
			Project plan and scope changes	Productivity analysis	Summarized	Projects, Contractors, Time	NA	NA	Low
			Quality control compliance	Work measurement and percent completion	Summarized	Projects, Contracts, Time	Summarized	Projects, Contracts, Time	High
				Change orders archive	Summarized	Projects, Contracts, Time	Detailed	Projects	Low
				Field inspection reports archive	Summarized	Projects, Time	Detailed	Projects	Medium
				Quality assurance checklist archive	Summarized	Projects, Contractors, WBS level	Detailed	Projects	Medium

118

Table 5.9: Decision-Support Requirements Matrix for Functional Management (Project Control Division)

Decision Level	Decision Node	Section	Decisions	Information Needs	Desired Granularity Level	Desired Dimensions	Available Granularity Level	Available Dimensions	Satisfaction Level (High/Med/Low)
Functional or Operational	Manager (Project Control Division)	Cost and Scheduling	Schedule control	Schedule variance analysis	Detailed, Summarized	Project, Time	Detailed	Project, Time	Medium to High
			Cost control	Cost variance analysis	Detailed, Summarized	Project, Time	Detailed	Project, Time	Medium to High
			Project and activities percent completion	Project budget	Summarized	Contracts, Time	Summarized	Contracts, Time	High
			Weekly projects progress evaluation	Work measurement and percent completion	Detailed	Projects, Contracts, Time	Detailed	Projects, Contracts, Time	High
		Contract Services	Contracts monitoring	Field inspection reports archive	Summarized	Projects, Time	Detailed	Projects	Medium
			Procurement management	Cost and schedule variance analysis	Summarized	Projects, Contracts, Time	Detailed	Projects	Medium
			Claims evaluation and recommendations	Change orders archive	Summarized	Projects, Contracts, Time	Detailed	Projects	Low
				Work measurement and percent completion	Detailed	Projects, Contracts, Time	Detailed	Projects, Contracts, Time	High
				Past project archive	Summarized	Projects, Claims	Detailed	Projects, Claims	Medium

119

Table 5.10: Decision-Support Requirements Matrix for Functional Management (Construction Division)

Decision Level	Decision Node	Section	Decisions	Information Needs	Desired		Available		Satisfaction Level (High/Med/Low)
					Granularity Level	Dimensions	Granularity Level	Dimensions	
Functional or Operational	Manager (Construction Division)	Construction Planning	Project economic feasibility	Economical analysis	Detailed, Summarized	Projects, Time, Size, Location	Detailed	Projects, Size	Medium
			Project budget	Financial analysis	Detailed, Summarized	Projects, Accounts, Time	Summarized	Projects, Accounts, Time	Medium to High
				Past project archive	Summarized	Cost, Size, Time	Summarized	Cost	Medium
		Construction Management Services	Contracts evaluation	Field inspection reports archive	Summarized	Projects, Time	Detailed	Projects	Medium
			Change order approvals	Work measurement and percent completion	Summarized	Projects, Contracts, Time	Summarized	Projects, Contracts, Time	High
				Project budget	Summarized	Contracts, Time	Summarized	Contracts, Time	High
				Cost and schedule variance analysis	Detailed, Summarized	Projects, Contracts, Time	Detailed	Projects	Medium

120

5.7 Summary

The first step in developing the decision-support framework was to understand the nature, work style and business processes of the construction owner organizations and to identify the user requirements for planning and decision-support operations. For this purpose, the modeling of selected owner organization, i.e. Miami-Dade Transit (MDT) was performed from different perspectives namely organizational, functional, informational, and decision. The organizational perspective illustrated the organizational structure and management hierarchies. The functional perspective showed how different business processes were executed and also depicted different business rules. The informational perspective represented data generation and storage in the organization. The decision perspective helped to identify the different decisions made at different management levels and information requirements for these decisions. These four models assisted to capture the existing or "AS-IS" state of MDT. In the next step, data and information flow models were prepared to identify the flow of operational data into, within and outside the organization. The data and information flow diagrams were mapped with the decision-perspective model to prepare decision-support matrices for different management levels. These matrices were used to identify the potential subject areas which were considered in the development of the decision-support framework. Although all models were developed particularly for MDT but they are generic to most construction owner organizations and can be used with little modifications.

CHAPTER 6

DEVELOPMENT OF FUNCTIONAL MODEL AND

REFERENCE ARCHITECTURE OF DATA WAREHOUSE

6.1 Introduction

In this chapter, the functional model and reference architecture of the proposed data warehouse are presented. The functional model represents how the system would work while the reference architecture reveals its physical and technical implementation using different software and hardware tools. Figure 6.1 illustrates the framework adopted for developing the data warehouse via the Bottom-up approach. Using this approach, first the data marts of different functional areas were developed and then integrated into an enterprise-wide data warehouse. More details about this approach are available in section 2.4.4.2 of Chapter 2. Each step shown in Figure 6.1 is fully explained in the following sections with technical details available in Appendix B.

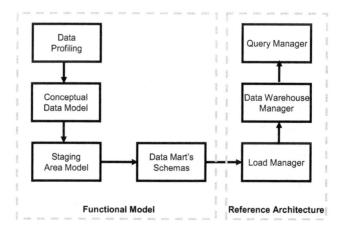

Figure 6.1: Framework adopted for Developing the Data Warehousing System

6.2 Data Profiling

Data profiling represents a systematic examination of the contents, structure and quality of a data source. At one extreme, a very clean data source that has been well maintained requires minimal transformation and human intervention to load directly into the final Dimension tables and Fact tables. At the furthest extreme, if data profiling reveals that the source data is deeply flawed and cannot support the business objectives, the data warehousing effort may need to be called off (Kimball and Caserta, 2004).

Three steps were performed to prepare the source data profiles:

(1) Identification of data sources within and outside the organization.

(2) Recognition of "data of interest" or useful data using decision-support matrices as shown in Tables 5.6-5.10.

(3) Examination of data format, syntax and units.

For this purpose, existing information systems at MDT were thoroughly studied to understand the data syntax and storage formats. The resulting source data profile is shown in Table 6.1. An examination of Table 6.1 indicates that most of the existing data were stored in the spreadsheet format. Since the spreadsheet data does not contain any data definitions, old project reports were also examined to identify the data types and corresponding units. This step was necessary to identify the similar data types which could later be integrated in the data warehouse. A preliminary check was performed to analyze the data quality by looking for any (1) missing values and (2) values with inconsistent data types. The results indicated that most of the data were in good shape and did not require an extensive cleaning effort.

Table 6.1: Source Data Profile

Classification	Description	Type	Unit	Source	Format*
Project Information	Project ID, type, location, size, duration, cost, project manager, funding agency, facility user, start and completion dates	Text	NA	Project information file	.xls
				Projects database	.mdb
Design	Preliminary and detailed project design	Drawing	NA	Project drawings	.dwg
	Design changes	Text, Drawing	NA	Design events report Revised drawings	.xls .dwg
Cost and Scheduling	Activity ID, Activity name, Activity type, Activity Duration, Activity times (EST, EFT, LST, LFT, TT, FF)	Numeric, Text	Days	Primavera scheduling files	.xls^
	Earned Value Parameters (BCWS, ACWP, BCWP, CV, SV, CPI, BAC, ETC, EAC)	Currency, Numeric	$	Earned value analysis file	.xls
	Structural and nonstructural items quantities, Material quantities	Numeric	ft^2, ft^3	Estimating files	.mdb^
	Project man hours (labor, skilled, engineers, managers etc.)	Numeric	hours	Estimating files	.mdb^
	Project Cost (cost per item, material cost, labor cost, equipment cost, overheads, miscellaneous costs, total cost)	Currency	$, $/ft^2$, $/ft^3$	Estimating files	.mdb^
Project Progress	Quantities of works done	Numeric	ft^2, ft^3	Daily project report	.xls
	Amount of works done	Currency	$	Daily project report	.xls
	Man hours used (full time and part time) per activity, time delay due to equipment, material and labor	Numeric	hours	Daily project report	.xls
	Information about Change orders and RFIs	Text	NA	Daily project report	.xls
	Revised project schedule, revised project cost	Numeric, Currency	Days, $	Schedule and cost control file	.xls

*.mdb: MS Access™ file, .xls: MS Excel™ file, .doc: MS Word™, .dwg: AutoCAD™ drawing
^ the softwares offer the facility to convert file from original program format to Access™ or Excel™ format

Table 6.1: Source Data Profile (continue)

Classification	Description	Type	Unit	Source	Format*
Financial	Cash flow sheets (contract amount, total estimated cost at completion, actual costs to date, total billed)	Currency, Numeric	$	Daily Cash flow sheet	.xls
	Budget files (Baseline budget, ceiling, percent budget spent, planned payments, actual payments, planned difference)	Numeric	$	Daily budget sheet	.xls
	Accounts balance sheets (fund type, fund ceiling, available funds, funds expiration date)	Currency, Numeric	$	Account balance sheets	.xls
Safety	Site safety plan details, Safety procedures in use, Insurance details	Text	NA	Site safety plan	.doc
	Site accident details, number of fatalities and injuries	Text	NA	Daily project report / Daily field inspection report	.xls / .xls
Quality	Quality checklist	Text	NA	Monthly quality report	.xls
	Quality related disputes	Text	NA	Field inspection reports	.doc
Contracts	Contract information (name, type, description, amount, duration)	Text	NA	Contracts database	.mdb
	Contractor's information (name, type, address, phone, fax, e-mail, ranking, prequalification information)	Text	NA	Contractors information file	.xls
	Permits information (name, agency, cost, status)	Text	NA	Permits information file	.xls
Claims	Claims information (name, type, amount, status, dates)	Text	NA	Claims database	.mdb
	Change orders information, RFI details, NTP information, Cost and schedule changes	Text	NA	Daily project report	.xls
	Project disputes information	Text	NA	Daily field inspection report	.doc

* .mdb: MS Access™ file, .xls: MS Excel™ file, .doc: MS Word™

125

6.3 Conceptual Data Model

The conceptual data model was prepared with two objectives in mind:

(1) To identify the subject areas for which data marts would be developed.

(2) To identify facts and dimensions for developing the data marts schemas.

Subject areas refer to the areas of interest or business processes about which users want to analyze the data or seek aid in decision-making such as productivity analysis, financial management, etc. (Ballard et al., 1998). For identifying the subject areas for MDT, a preliminary list was prepared using the decision-support matrices (Tables 5.6-5.10). This list was refined using feedback from the MDT's staff and subject areas were ranked according to the organization needs. The final list of subject areas with their brief description is shown in Table 6.2.

Table 6.2: Identification of Subject Areas

No.	Subject Area Title	Description
1.	Project Performance	Project performance analysis in terms of cost, schedule, completion, safety and claims. Users could compare the performance of different projects with reference to time, location, type of project and contactors.
2.	Productivity Information	Project productivity analysis and evaluation of productivity index in different projects with respect to time, different contractors and different project conditions.
3.	Contractor Performance	Contractor's performance in different projects in terms of budget, schedule, safety and quality. Users could compare the performance of different contractors on the same type of projects or performance of one contractor on different types of projects.
4.	Schedule Management	Project schedule information and its analysis with reference to different projects, contracts, contractors and project activities. It will also analyze the effect of change orders on project schedule.
5.	Cost Management	Project cost information and its analysis with reference to different projects, contracts, contractors, project activities and change orders.
6.	Financial Management	Financial performance of Miami-Dade Transit (MDT) on various projects with reference to time, contracts and contractors.
7.	Claims Management	Claims information with reference to different projects, time, contracts and contractors.
8.	Safety Management	Project safety analysis with reference to different projects, time, contracts and contractors.

The next step involved in preparing the conceptual data model is to identify facts and dimensions. Facts represent quantitative (or factual) data about a business entity/transaction (an entity is an object or event for which we need to capture and store data) while dimensions contain descriptive data that reflect the dimensions of that entity. In other words, fact data contains the physical information about a factual event (e.g. *Cost, Budget, Measured Productivity*, etc.) and the dimension data shows the description of that fact (e.g. *Time, Location, Project,* etc.). The fact data is more stable than the dimension data, meaning that the dimension data change more frequently over a period of time than the fact data (Gray and Watson, 1998).

To identify facts and dimensions, the *functional* and *informational* models of MDT (as discussed in sections 5.4.2 and 5.4.3) were carefully examined. Initially, a list of all facts and dimensions was prepared. In the next step, these facts and dimensions were categorized according to the subject areas. The facts or dimensions which could not correspond to the existing subject areas were considered as *redundant* or not useful within the current scope of the data warehouse. A list of final facts and dimensions is shown in Table 6.3. It is important to note that the dimension *Time* had been represented in two hierarchies, *Real Time* hierarchy and *Fiscal Time hierarchy*. In the former one, time was organized as real date, month and year. While in the later one, time was arranged in fiscal days, weeks, months, and years. Such an arrangement was necessary as the executives required both types of hierarchies to analyze and/or compare different projects.

Table 6.3: Identification of Facts and Dimensions

Subject Area	Main Facts (Measures)	Dimensions								
		Time	Location	Project	Contract	Contractor	Project Conditions	WBS	Change Orders	Accounts
Project Performance	Cost, Schedule, Safety, Claims	*Real Time*	Site ID	Project ID	Contract-ID	Contractor-ID	Site-Conditions	Item Code	Change Order-ID	Account-ID
		Date	Street	Name	Type	Name	Weather-Conditions	Section	Category	Type
		Month	City	Type	Value	Type	Hazardous-Conditions	Level		Name
		Year	Zip	Size	Duration	Specialty		Segment		
			Zone	Cost	Specialty	Crew Size				
			County	Start Date		Crew-Specialty				
		Fiscal Time		End Data						
		Day No		Consultant		*Labor*				
		Week No		Manager		Skilled				
		Month No				Non-skilled				
		Quarter				Full Time				
		Half				Part Time				
		Year No								
						Address				
						Street				
						City				
						Zip				
						Phone				
						Fax				
						Email				

6.4 Staging Area Model

The purpose of staging area model, sometimes referred as the ETL (Extract-Transform-Load) model, is to extract data from the source systems, transform different source data standards into a single data standard, enforce data quality and consistency standards, integrate data so that separate sources can be used together, and finally deliver target data which are stored in the data marts schemas (Kimball and Caserta, 2004). The entire operation is depicted in Figure 6.2.

Figure 6.2: Staging Area Modeling Steps

6.4.1 Data Extraction and Transformation

The data extraction and transformation scheme developed for MDT is shown in Figure 6.3.

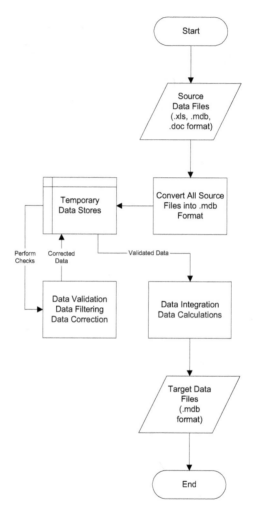

Figure 6.3: Flow Chart of Data Extraction and Transformation Scheme

As indicated in Table 6.1, the source data were available in different formats and need to be converted into one unique format. It was decided that the staging area prototype system would be developed in MS Access™ due to its simplicity and easy availability. Hence all data formats were converted into MS Access™ format (i.e. *.mdb* format). For the scope of this research, most of data extraction and transformation operations were performed manually. However, these operations could be automated by writing *Application Interface Protocols* (APIs) for different source systems. API is a computer program, commonly written in *Visual Basic for Applications (VBA™)*, and serves as an interface between the two main applications.

The transformed data were stored in the *temporary data files* for performing data quality and consistency checks and then for integrating the data. The final version of the data, called target data, was stored in the permanent database files. It is important to note that the source data stored in different information systems at MDT were not normalized and they were kept in the same format in the target data files.

6.4.2 Data Validation, Filtration and Integration

The purpose of data validation is to ensure data quality by correcting any errors, omissions, or inaccuracies before the data are loaded into the data marts. For data validation, the following checks were performed for each record in the source data files:

(1) Uniformity check to detect any out of range values or impossible values

(2) Conformity check to notice mismatching data units and data types

(3) Syntax check to identify erroneous field lengths

(4) Check for any missing values

The flow chart of data validation and integration scheme is shown in Figure 6.4. As can be seen in this figure, each faulty record was corrected before moving to the next record. To automate this operation, *macros* were written in SQL^{TM} and VBA^{TM}.

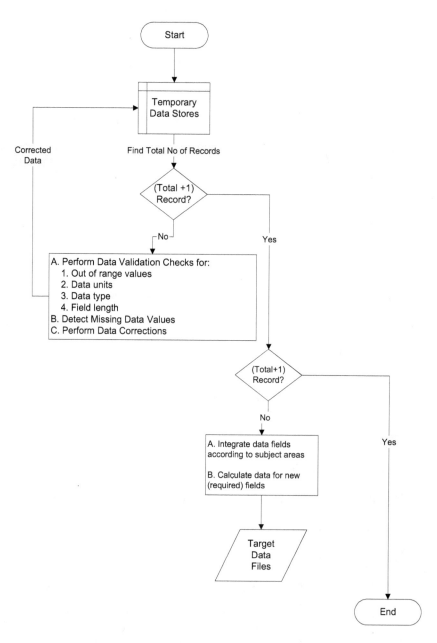

Figure 6.4: Flow Chart of Data Validation, Filtration and Integration Scheme
(Concept: Al-Deek, 2004)

A sample macro for checking a valid value in a column of a data table is shown below.

```
SELECT     unique_identifier_of_records
FROM       work_in_queue_table ProjInfo
WHERE      source_system_name = 'MDTSource_PjIn'
AND        column NOT EXISTS
           SELECT anything
           FROM   column_validity reference_table
           WHERE column_name = "Project_ID"
           AND    source_system_name = 'MDTSource_PjIn'
           AND    valid_column_value = ProjInfo.column_value
```

The data validation rules used to write these macros are summarized in Table 6.4. The table also indicates the corrective actions taken to correct any faulty or missing values. After data validation, data filtration was performed to delete duplicate data and rearrange the data according to the format of facts and dimensions tables.

After the data validation and filtration operations, data integration was performed. The purpose of data integration was to fuse data from different data sources according to the subject areas. Two types of integration operations were performed:

(1) Integration of similar data types from different sources

(2) Integration of different data types as per subject areas

As data in different sources could have different aggregation levels, a lowest aggregation level for each data type was adopted to capture the most detailed data. For example, for time-related data, all fields were stored in days. During the data integration process, some new data fields were also calculated to expedite the data processing during queries. The final data were stored in the permanent data files (in the .mdb format) which are referred as the target data. Table 6.5 explains the data integration rules as well as the target data profiling. The target data file IDs are shown with a suffix of "xxxx" which represents the last 4-digits of project IDs used by the MDT.

133

Table 6.4: Data Validation Rules

Data	Data Validation Checks				Corrective Actions
	Data Range/Domain	Correct Units	Data Type	Field Length	
Project ID	--	--	Text	Exactly 12	Ask user for corrections
Project Type	Road works, Bridge, Facility	--	Text	Max 20	Ask user for corrections
Project Location	Must be within Miami-Dade county	--	Text	Max 20	Match ID with Locations List
Project Size	No zero or negative values	ft, ft² or miles	Numeric	Max 10	Ask user for corrections
Project Duration	No zero or negative values	Days	Numeric	Max 4	Ask user for corrections
Project Cost	No zero or negative values	$	Currency	Max 20	Ask user for corrections
Project Personnel Names	--	--	Text	Max 30	Ask user for corrections
Activity ID	--	--	Text	Exactly 4	Ask user for corrections
Activity Type	--	--	Text	Max 10	Match ID with WBS Table
Activity Times	No negative values	Days	Numeric	Max 4	Recalculate the field
BCWS, ACWP, BCWP, BAC, ETC, EAC	No zero or negative values	$	Currency	Max 10	Recalculate the field
Cost Variance	--	$	Currency	Max 10	Recalculate the field
Schedule Variance	--	Days	Numeric	Max 10	Recalculate the field
CPI, SPI	No zero or negative values	--	Numeric	Max 10	Recalculate the field
Item or material quantities	No zero or negative values	ft², ft³	Numeric	Max 10	Ask user for corrections
Equipment & Man hours	No zero or negative values	Hours	Numeric	Max 10	Ask user for corrections
Project Costs, Claims Amount, Contract Value	No zero or negative values	$	Currency	Max 12	Recalculate the field
Project Completion	No zero or negative values, should be 100% or below	Percent	Numeric	Max 3	Recalculate the field
Budget amounts, Cash Flow Values, Funds	--	$	Currency	Max 12	Ask user for corrections
Fatalities and Injuries Ratios	No negative values	--	Numeric	Max 10	Ask user for corrections
All other IDs	--	--	Text	Max 12	Check with the valid IDs list
All other Names	--	--	Text	Max 20	Ask user for corrections

Table 6.5: Target Data Profile

Classification	Existing Fields	New Calculated Fields	Type	Conversion/Calculation Rules	File ID
Project Information	Project ID, Type, Location, Size, Duration, Cost, Project Manager, Funding Agency, Facility User, Start Date, Completion Date, Conditions	None	Text	Fields are extracted from 2 source files (.xls and .dmb format), rearranged and converted in .mdb format	ProjInfoxxxx
Design	Drawing ID, Description, Original Design Date, Design Approval Date, Revised Design Date, No of Revisions, Design Consultant	None	Text	.xls to .mdb conversion	Designxxxx
Scheduling	Activity ID, Activity Name, Activity Type, Activity Duration, Early Start, Late Start, Early Finish, Late Finish, Total Float, Free Float, BCWS, ACWP, BCWP, CV, SV, CPI, BAC, ETC, EAC	Project Delay, Cost Overrun	Numeric, Currency	.xls to .mdb conversion	Schedxxxx
Cost	Item ID, Description, Size, Material Quantity, Man Hours, Equipment Hours, Overheads, Unit Cost, Quantity, Total Cost	None	Numeric, Currency	None	Costxxxx
Project Progress	Item Code, Quantity of Work Done, Payment Amount, Change Orders, RFI, Percent Completion, Revised Project Cost, Revised Completion Date, Planned Productivity, Delays Information, Quality Measurement Index	Daily Productivity, Weekly Productivity, Cumulative Productivity, Productivity Performance Index	Numeric, Currency, Text	Productivity = Quantity of Work/Man Hours; Productivity Index = Actual productivity/planned productivity	PjPgxxxx

135

Table 6.5: Target Data Profile (Continue)

Classification	Existing Fields	New Calculated Fields	Type	Conversion/Calculation Rules	File ID
Financial	Account ID, Account Type, Account Name, Account Balance, Fund Type, Fund Ceiling, Funds Expiration Date	None	Currency	.xls to .mdb conversion	Fin1xxxx
	Project ID, Baseline Budget, Ceiling, Percent Budget Spent, Planned Payments, Actual Payments, Planned Difference	None	Currency	.xls to .mdb conversion	Fin2xxxx
Safety	Project ID, Accidents Count, Injuries Count, Fatalities Count, Accident Claim Amount, Insurance Type, Insurance Coverage	Injuries Rate, Fatalities Rate	Text	Fields are extracted from 2 source files (.xls and .doc format), rearranged and converted in .mdb format	Safexxxx
Contract	Contract ID, Type, Description, Amount, Duration, Specialty, Permit Name, Permit Status	None	Text	None	Contxxxx
Contractor	Contractor ID, Name, Type, Specialty, Street Address, City, State, Zip, Phone, Fax, Email, Prequalification Information	None	Text	None	Contrxxxx
Claims	Claim ID, Claim Count, Claim Type, Claim Amount, Parties, Submission Date, Resolution Date, Status, Settled Payment	None	Text	None	Claimxxxx

6.5 Data Marts Schemas

A data mart schema is a *dimensional model* which is composed of a central *fact table* and a set of surrounding *dimension tables*. The dimension tables are connected with the fact table by foreign keys (a foreign key is a primary key of one entity that is also an attribute in another entity). As a result, a fact table contains facts (also called measures) and foreign keys to the dimension tables (Levene and Loizue, 2003).

There are three main types of data mart design schemas: the *star schema*, the *snowflake schema* and a hybrid of the two, the *starflake schema*. The star schema is the simplest database structure containing a fact table in the center which is surrounded by the dimension tables. The star schema uses denormalized data to provide fast response times, allowing database optimizers to work with simple database structures in order to yield better execution plans. The snowflake schema is a variation of the Star structure, in which all dimensional information is stored in the third normal form, while keeping fact table structure the same. This schema is good in situations where the operational data are stored in the third normal form. The starflake schema is a combination of the denormalized star schema and the normalized snowflake schema. It is used in situations where it is difficult to restructure all entities into a set of distinct dimensions (Ahmad et al., 2004).

Since most the operational data of MDT were available in the denormalized format, it was decided to adopt the Star schema for designing the data marts of each subject area. The Star schemas of two subject areas namely *Project Performance* and *Productivity Information* are shown in Figures 6.5 and 6.6 while the schemas of remaining subject areas are depicted in Appendix B.

137

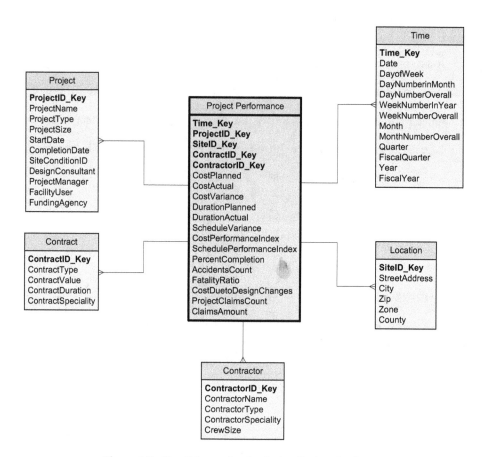

Figure 6.5: Star Schema for Analyzing Project Performance

As can be seen in Figure 6.5, "Project Performance" is the fact table which is surrounded by five dimension tables namely "Project", "Contract", "Contractor", "Time" and "Location". These dimension tables are connected to the fact table through foreign keys (FKs). Each bold face attribute with a suffix "_key" represents a FK. The cardinality relationships between fact table and dimensions tables are "many-to-one" to ensure their hierarchy.

138

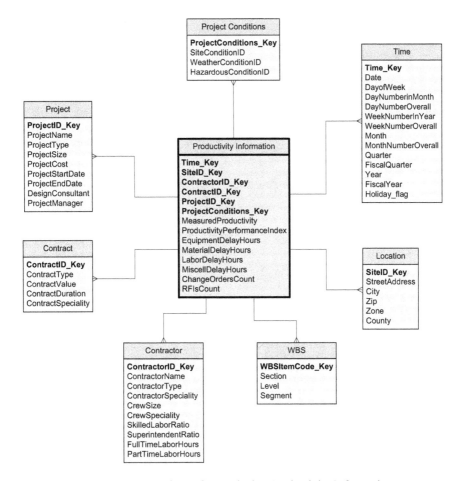

Figure 6.6: Star Schema for Analyzing Productivity Information

6.5.1 Metadata

Metadata is defined as data about data. It is like a card index describing how

information is structured within the data mart schemas, i.e. it defines field names, their

data type, field sizes, data format, any validation rules and other such attributes.

Metadata is used for a variety of purposes. As part of the data extraction and load

process, it is used to map data sources to the common view of information within the data mart. As part of the query management process, metadata is used to direct a query to the most appropriate data source (Krishna, 2004). The metadata of data mart schemas is shown in Table 6.6 while the metadata of fact and dimension tables is depicted in Appendix B (Tables B1 and B2). The direct users as indicated in Table 6.6 are MDT executives who would have authorization to make any changes in the data mart schemas.

Table 6.6: Metadata of Data Mart Schemas

No.	Name	Primary Key	Aggregation	Load Frequency	Direct User(s)
1.	Project Performance	Combination of ProjectID and LocationID keys	Weekly, Monthly	Weekly	Deputy Director, P&D Assistant Director, C&E Chief, Project Control Chief, Construction
2.	Productivity Information	Combination of ProjectID, ContractorID and LocationID keys	Weekly, Monthly	Weekly	Chief, Project Control
3.	Contractor Performance	Combination of ContractorID and ProjectID keys	Weekly, Monthly	Weekly	Chief, Construction Manager, Contract Administration
4.	Schedule Management	Combination of WBSItemCode and ProjectID keys	Weekly, Monthly	Weekly	Chief, Construction Manager, Cost and Scheduling
5.	Cost Management	Combination of WBSItemCode, ProjectID and ContractID keys	Weekly, Monthly	Weekly	Chief, Construction Manager, Cost and Scheduling
6.	Financial Management	Combination of AccountsID and ProjectID keys	Weekly, Monthly, Yearly	Monthly	Chief, Project Control Chief, Construction
7.	Claims Management	Combination of Time, ProjectID and ContractID keys	Monthly	Monthly	Manager, Claims Administration
8.	Safety Management	Combination of Time, Location and ProjectID keys	Monthly	Monthly	Chief, Construction Manager, Contract Administration Manager, Claims Administration

6.6 Load Manager

The load manager is used to load target data into the respective facts and dimension tables of the data marts. It performs two operations:

(1) It map target data with the fields of fact tables and dimensions tables.

(2) It consolidates facts and dimensions data according to the selected granularity level.

Table 6.7 illustrates the mapping process of target data and data marts. It shows the target data files and their respective data fields which were used to populate the respective fact tables and dimension tables of several data marts. Several commercial off-the-shelf softwares are available which can be used as a load manager such as SAP R/3™, Carelton Passport™, etc. As the emphasis of this research was not on developing a software product, this operation was performed partially manually and partially using *macros* which were written in MS Access™.

Granularity refers to the level of detail provided by a data record in a data mart. The more detail, the lower the level of granularity. Granularity allows users to drill down into details and roll up into the aggregated groups. The choice of granularity requires trading off volume of data for level of query detail. A second-trade-off is between the level of granularity and the amount of computing power required. Table 6.8 shows the granularity levels suggested for the *Time*, *Location* and *WBS* dimensions. For developing the prototype system, the lowest level of granularity was chosen due to the small data volume.

Table 6.7: Target Data to Data Mart Mappings

Name	Data Mart Data Type	Data Mart Mappings		
		Measures/Attributes	Target Data Fields	Target Data File ID
Project Performance	Fact	Planned Cost, Actual Cost, Cost Variance, Schedule Variance, Cost Performance Index, Schedule Performance Index	BCWS, ACWP, BCWP, CV, SV, CPI, SPI	Schedxxxx
		Percent Completion, Planned Duration, Actual Duration, Cost Changes due to Change Orders	Project Duration, Percent Completion, Revised Project Cost, Revised Completion Date	PjPgxxxx
		Accidents Count, Fatality Count	Accidents Count, Fatalities Count	Safexxxx
		Project Claims Count, Claims Amount	Claim Count, Claim Amount	Claimxxxx
Productivity Information	Fact	Measured Productivity, Productivity Performance Index, Equipment Delay Hours, Material Delay Hours, Labor Delay Hours, Miscellaneous Delay Hours, Change Orders Count, RFIs Count	Daily Productivity, Weekly Productivity, Cumulative Productivity, Productivity Performance Index, Change Orders, RFIs, Delays Information	PjPgxxxx
Contractor Performance	Fact	Cost Performance Index, Schedule Performance Index	CPI, SPI	Schedxxxx
		Productivity Performance Index, Equipment Delay Hours, Material Delay Hours, Labor Delay Hours, Quality Measurement Index	Productivity Performance Index, Delays Information, Quality Measurement Index	PjPgxxxx
		Site Accidents Count	Accidents Count	Safexxxx
Schedule Management	Fact	Planned Period, Actual Period, Planned Start Time, Actual Start Time, Planned Finish Time, Actual Finish Time, Schedule Variance, Schedule Performance Index	Activity ID, Activity Duration, Early Start, Late Start, Early Finish, Late Finish, Total Float, Free Float, SV, SPI	Schedxxxx
Cost Management	Fact	Planned Cost, Projected Cost, Actual Cost, Cost Variance, Cost Performance Index	BCWS, ACWP, BCWP, CV, CPI, BAC, ETC, EAC	Schedxxxx Costxxxx

142

Table 6.7: Target Data to Data Mart Mappings (Continue)

Name	Data Mart Data Type	Data Mart Mappings		
		Measures/Attributes	Target Data Fields	Target Data File ID
Financial Management	Fact	Base Line Budget, Budget Ceiling, Charges to Date, Planned Payment, Actual Payment, Planned Difference, Overheads, Available Funds Percent Budget Spent	Account ID, Account Balance, Baseline Budget, Ceiling, Percent Budget Spent, Planned Payments, Actual Payments	Fin1xxxx Fin2xxxx
Claims Management	Fact	Claim Type, Claim Amount, Claim Parties, Claim Status, Claim Submission Date, Claim Resolution Date, Settled Payment	Claim ID, Claim Count, Claim Type, Claim Amount, Parties, Submission Date, Resolution Date, Status, Settled Payment	Claimxxxx
Safety Management	Fact	Accident Type, Injuries Count, Fatalities Count, Injuries Rate, Fatalities Rate, Accident Claim Amount, Insurance Type, Insurance Coverage	Project ID, Accidents Count, Injuries Count, Fatalities Count, Accident Claim Amount, Insurance Type, Insurance Coverage	Safexxxx
Time	Dimension	Date, Day of Week, Day Number in Month, Day Number Overall, Week Number in Year, Week Number Overall, Month, Month Number Overall, Quarter, Fiscal Quarter, Half, Year, Fiscal Year, Holiday Flag	Date, Day, Week, Month and Year fields in each target file	ProjInfoxxxx Schedxxxx PjPgxxxx Safexxxx Claimxxxx Fin1xxxx Fin2xxxx Costxxxx
Location	Dimension	Site ID, Street Address, City, Zip, Zone, County	Location (ID, Street, City, Zip, Zone, County)	ProjInfoxxxx
Project	Dimension	Project ID, Project Name, Project Type, Project Size, Project Cost, Project Start Date, Project End Date, Design Consultant, Project Manager	Project ID, Type, Size, Duration, Cost, Project Manager, Funding Agency, Facility User, Start Date, Completion Date	ProjInfoxxxx
Contract	Dimension	Contract ID, Contract Type, Contract Value, Contract Duration, Contract Specialty	Contract ID, Type, Description, Amount, Duration, Specialty, Permit Name, Permit Status	Contxxxx

Table 6.7: Target Data to Data Mart Mappings (Continue)

Name	Data Mart Data Type	Data Mart Mappings		
		Measures/Attributes	Target Data Fields	Target Data File ID
Contractor	Dimension	Contractor ID, Contractor Name, Contractor Type, Contractor Specialty, Crew Size, Crew Specialty, Skilled Labor Ratio, Superintendent Ratio, Full Time Labor Hours, Part Time Labor Hours, Address	Contractor ID, Name, Type, Specialty, Street Address, City, State, Zip, Phone, Fax, Email, Prequalification Information	Contrxxxx
Project Conditions	Dimension	Site Conditions ID, Weathers Condition ID, Hazardous Conditions ID	Project Conditions	ProjInfoxx xx
WBS	Dimension	WBS Item Code, Section, Level, Segment	Activity ID, Item ID	Schedxxxx Costxxxx
Change Orders	Dimension	Change Order ID, Change Order Category	Change Orders	PjPgxxxx
Accounts	Dimension	Account ID, Account Type, Account Name	Account ID, Account Type, Account Name	Fin1xxxx

Table 6.8: Granularity Levels for TIME, LOCATION and WBS Dimensions

Granularity Level	Dimensions		
	Time	Location	WBS
	Week	City	Section
	Month	County	Level
	Year		Segment

6.7 Warehouse Manager

The functions of warehouse manager are to store and manage data and their corresponding metadata in the data marts, to archive or purge old data and to prepare data summaries. The metadata contains the rules for data archiving and summarization. Different commercial off-the-shelf and customized software products are available for building the warehouse manager. In this research, the warehouse manager was designed in *MS Access ™*. More details about the warehouse manager are available in section 6.9.

144

6.8 Query Manager

The query manager performs all operations related with the query management process and also serves as a back-end for the decision-support applications. In this research, *Power OLAP®* was used for building the query manager. It is an *MS Excel™ Add-in* which could perform necessary OLAP operations such as slice and dice, roll-up, roll-down and roll-across the data.

6.9 Development of Prototype System

A prototype system was developed to verify the functional model and to demonstrate the capabilities of the proposed data warehouse to the MDT executives. The reference architecture of the prototype system is shown in Figure 6.7. The load manager and warehouse manager were developed using *MS Access™*. The query manager was designed using *Power OLAP™* and *MS Excel™*. The client-end was developed in *Xcelsius™* which is a powerful tool for generating interactive graphs and tables.

As can be seen in Figure 6.8, the load manager extracts operational data stored in spreadsheets and applications databases (such as Primavera®, Timberline®, etc.) and transform them into *MS Access™* format. Macros were written in *MS Access™* to validate, filter and integrate the data according to the subject areas of data marts. The final clean data were stored in their corresponding data marts developed in *MS Access™*. The user can retrieve data through an interface designed in *Power OLAP®* and then perform different OLAP operations. The *Power OLAP®* exports selected data to *MS Excel®* and *Xcelsius™* for preparation of charts and graphs. The selected screenshots of the prototype system are shown in Appendix C.

145

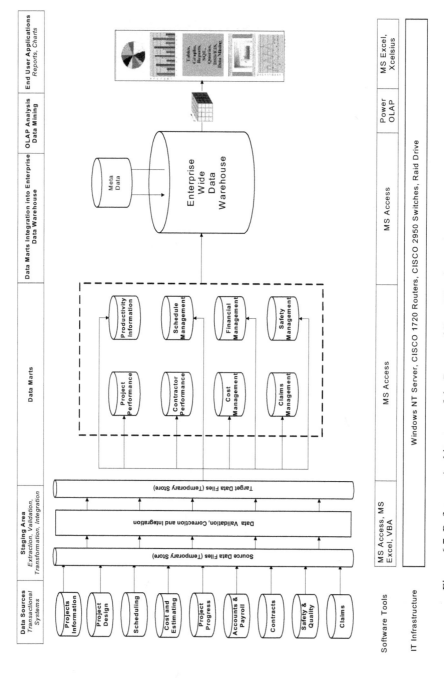

Figure 6.7: Reference Architecture of the Proposed Data Warehouse for MDT (Concept: Furno, 2003)

146

6.9.1 Verification of the Prototype System

For verification purposes, limited data were fed in the prototype system. The results were compared with the existing project reports. It was found that the prototype system results were same as those prepared by using manual calculations or customized spreadsheets. This comparison of results verified that the logics and schemas used in the development of the prototype system were correct. Figure 6.8 illustrates a set of results obtained from the "Project Performance" data mart. These results demonstrate that via data warehousing users could analyze data from different perspectives and explore it up to the desired level of detail which is often required in different planning and decision-making operations. A sample run of the prototype system along with detailed demonstration results is given in Appendix C (All these results are prepared for demonstration purposes only and do not reflect the actual project data).

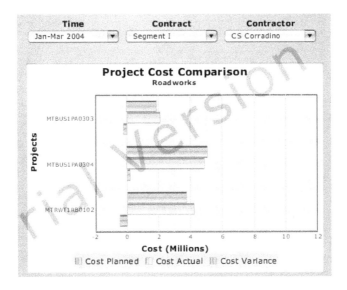

(a): Project Cost Comparison for January-March 2004 Quarter of Segment I

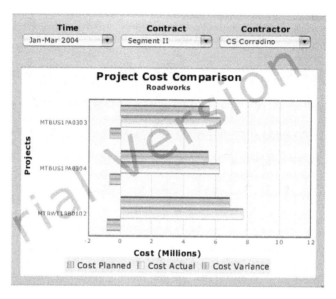

(b): Project Cost Comparison for January-March 2004 Quarter of Segment II

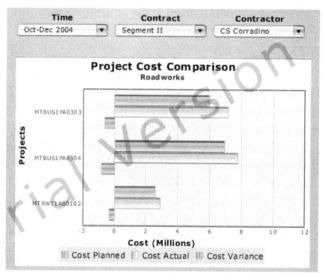

(c): Project Cost Comparison for October-December 2004 Quarter of Segment II

Figure 6.8: Project Cost Comparison Using Multidimensional Analysis

6.10 Summary

Based on user requirements and decision needs of the collaborator organization, a functional model of the proposed data warehouse was prepared using the Bottom-up approach. The functional model was composed of four components namely data profiling, preliminary data model, staging area model and data marts schemas. The purpose of data profiling was to identify source data stored in different operational systems and to examine their format, syntax and data type. Then the preliminary data model was prepared which showed different facts and dimensions categorized according to different subject areas or data marts. The staging area model contained several interrelated models to extract, validate, filter and integrate data. This step was extremely important to ensure data quality. At the end, data marts schemas were prepared using the star structure.

The reference architecture was composed of a load manager, warehouse manager and a query manager. Prototype system was developed using *MS Access ™, Power OLAP ™ MS Excel ™* and *Xcelsius ™* to demonstrate the capabilities of the proposed data warehouse. Limited data were feed in the prototype software and results were compared with the existing project documents for verification. Several refinements were made to fine tune the prototype system.

CHAPTER 7

DEVELOPMENT OF THE DECISION-SUPPORT FRAMEWORK

7.1 Introduction

The proposed decision-support framework is presented and discussed in this chapter.

During the course of this research, it is found that construction owner organizations

typically have a hierarchical organizational structure. In organizations with a hierarchical

structure, the responsibility of intermediate organizational units is to process data and

information to prepare reports to be used by executives for making short and long term

decisions. A successfully implemented data warehouse can help carry out this

responsibility with increased effectiveness and efficiency and as a result, many of these

intermediate organizational units may no longer be needed. The implication is that

implementation of data warehousing could lead to major restructuring of organizations.

Watson et al. (2001) analyzed different organizations that have implemented data

warehousing during the last three to five years. They concluded that the organizations

which redesigned their strategies and restructured their organizations for effective

implementation of data warehousing earned more monetary benefits than the

organizations that chose not to alter their organizational structures.

Besides, the information technology (IT) infrastructure of most organizations is based

on the traditional *applications-centric* approach where the data are stored and shared

through different discrete applications. Data warehousing offers a new concept of *data-*

centric approach where the data are stored and shared through a central repository. As a

result, the IT infrastructure may also need to be redesigned before the implementation of

a data warehouse. Hence it was decided that the proposed decision-support framework

150

would be composed of three phases namely, data warehousing model (already presented in chapter 6), IT infrastructure model and an organizational restructuring model as shown in Figure 7.1. A new IT infrastructure and a new organizational structure for the Miami-Dade Transit (MDT) are proposed in this chapter. It should be pointed out however, that the IT infrastructure and the organizational structure presented are generic in nature, and can be implemented, in principle, in other public owner organizations with some modifications. The proposed decision-support framework was verified and validated first by the MDT and then by twenty large construction owner organizations (chosen from the respondents of the survey mentioned in chapter 4) within the continental United States.

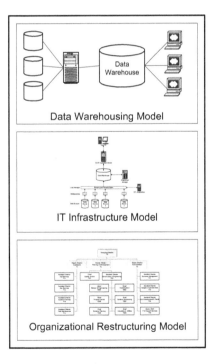

Figure 7.1: Components of the Proposed Decision-Support Framework

7.2 Information Technology (IT) Infrastructure Model

The IT infrastructure at MDT, like in many other construction owner organizations, has not been planned and developed along with eventual "computerization" of different units and processes during the last two to three decades. This IT infrastructure can be termed *application-centric* meaning that separate applications are developed to run various discrete processes and as a result, data needed by individual processes are stored in individual applications. There is minimal integration among these applications, and users often have to manually integrate data from different applications to generate information often needed for the purpose of decision-making. Kim (2001) defined four levels of integration between different software applications: *connectivity*, *sharing*, *interoperability* and *coordination*. Connectivity is the minimum level of integration where the two software applications (or information systems) are linked via a network while coordination represents the highest level where data from different applications are integrated and shared. At present, the integration level at MDT is even below the connectivity stage (as all applications are not linked via a network) as shown in Figure 7.2. Although various organizational divisions at MDT are linked via a Local Area Network (LAN), but this network is mainly used for internet and intranet functions. Data sharing among many different users and applications are done either via removable storage devices (floppy disk, CD, flash disk etc.) or via electronic file exchanges (e-mail attachments or electronic file transfer). This implies that at any given time, different users may have using different versions of the same data set.

For implementing data warehousing which is based on *data-centric* approach, the current IT infrastructure needs to be redesigned to allow real time two-way communications between the users (or data sources) and the data warehouse.

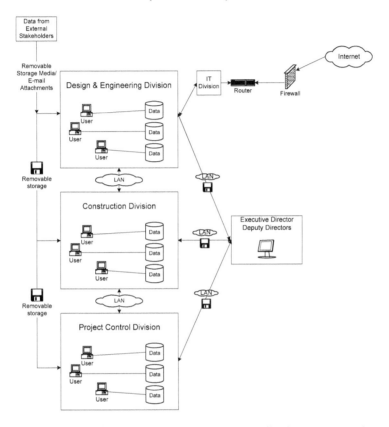

Figure 7.2: Existing IT Infrastructure at MDT (Palisades Group, 2004)

The proposed IT infrastructure which is based on the Center-to-Center (C2C) communications concept is shown in Figure 7.3. Center-to-Center (C2C) is a two-way communication approach which means that all units in an organization are interconnected via a network and periodically supply data to the data warehouse which stores it in a central repository (Courage et al., 2004). These units can retrieve the same information from the data warehouse at any real time. The individual units can be granted different levels of access for information retrieval to ensure data safety and privacy. This IT infrastructure was developed in consultation with the IT division of MDT.

Presently, data from various construction sites as well as from different stakeholders are first supplied to the respective MDT divisions, which load the data in different applications or information systems. By implementing the new IT infrastructure, these data can be directly transferred to the data warehouse via wire line or wireless communications. It will result in substantial time savings and will minimize any chances of data corruption or vandalism. Besides making implementation of data warehousing, simpler, this new IT infrastructure will increase coordination between various information systems as well as business units. For example, data from different construction sites can be accessed in real time thus enabling critical decisions to be made instantly.

The data warehouse may be implemented in the form of a one-tier (monolithic system), two-tier (client/fat server) or three-tier (client/slim servers) architecture. One-tier system is appropriate for a single data mart and hence not considered any further. In a two-tier architecture, business logic or decision-support engine, data management, and data presentation functions reside on the "client", traditionally the user's workstation. In a three-tier architecture, any or all of these functions reside on intermediate servers (Dyché, 2000). In this research a three-tier architecture was adopted due to its flexibility, increased processing speed and high level of security. With three-tier architecture, the decision-support applications are separate from the main data warehouse; thus allowing easy interface to the users without interfering with the data management operations of the data warehouse. Figure 7.4 demonstrates the reference architecture for the data warehouse implementation using a three-tier architecture.

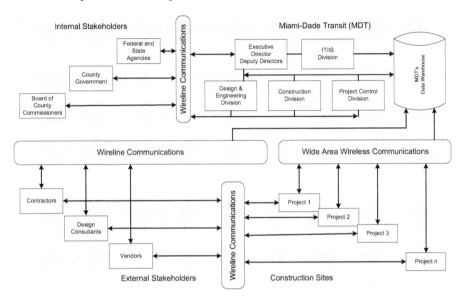

Figure 7.3: Proposed IT Infrastructure for MDT (Concept: Courage et al., 2004)

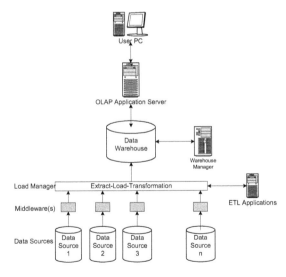

Figure 7.4: Reference Architecture of the Data Warehouse using Three-Tier Architecture

155

As indicated in Figure 7.4, the middleware are software tools that allow discrete information systems or applications to share data. There are different types and categories of middleware depending on system needs and user requirements. After consultation with IT department of MDT, it was decided to adopt *Application Interface Protocol* (API) approach for designing the middleware(s). API is a fixed set of procedures written in any standard programming language, callable from an external user-generated program, which provides access to some of the functions of the application system or to the data. Further, APIs have different design architectures and it is recommended in this study to use the *Common Object Request Broker Architecture* (CORBA). CORBA is an object-oriented standard designed especially for distributed systems. The main advantages of CORBA are its reusability, platform and language independence. CORBA not only enables systems to retrieve data from other systems, but also remotely activate methods embedded in objects. The network connection is peer-to-peer (Al-Deek and Abd-Elrahman, 2002).

The data warehousing implementation also requires appropriate modifications in the network to extract, load, manage and retrieve the data. Figure 7.5 illustrates the proposed network configuration for data warehousing. This configuration is based on the three-tier architecture as discussed earlier. For data security, two firewalls are proposed as suggested by Harmon (1998) to protect the network from breaches via Internet. The internal MDT users will have access to the data warehouse via local terminals while the external users will retrieve allowable information using a web browser. This requires a web interface for the proposed data warehouse which involves development of a commercial software product and is considered beyond the scope of this research study.

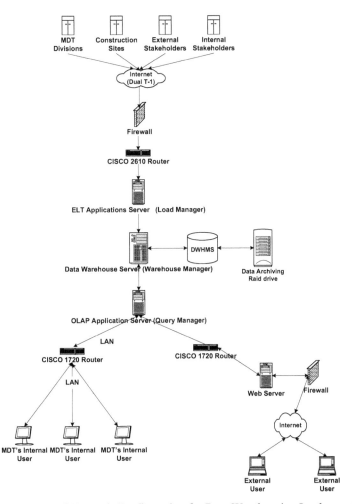

Figure 7.5: Proposed Network Configuration for Data Warehousing Implementation
(Concept: Courage et al., 2004)

The description of different software and hardware components along with their costs is given in Table 7.1. The selection of software tools and hardware components was done inline with the MDT's IT strategic plan (Palisades Group, 2004). The software and hardware costs are obtained from quotes given by the respective distributors in May-June 2005 and are subject to change.

Table 7.1: Estimated Hardware and Software Costs for Data Warehousing Implementation

Item	Quantity	Unit Price ($)	Total Price ($)
Hardware Items			
Dell Power Edge SC430 servers	3	25,000	75,000
Dell 6850 database server	1	18,000	18,000
Western Digital 400 GB raid drive	1	8,000	8,000
Routers and interface cards for T1 connections	4	4,000	16,000
Application development workstations	2	6,000	12,000
Network switches and other miscellaneous costs			10,000
Total Hardware Cost			*139,000*
Software Items			
DBMS software (Oracle® with 25 licenses)			30,000
OLAP Applications tools (Power OLAP®)			12,000
MS Windows Advanced Server			3,200
Miscellaneous system management tools			25,000
Total Software Costs			*70,200*
System Development Costs			
Project Manager	1	$80/hour	128,000
Database Administrator (DBA)	1	$75/hour	120,000
Senior Programmer	1	$75/hour	120,000
Programmers	2	$60/hour	96,000
Systems + Network administrator	1	$60/hour	96,000
Miscellaneous costs and overheads			50,000
Total System Development Costs			*610,000*
Miscellaneous			
Users training	50	$800/user	40,000
Total Costs for One Year			**859,200**
Operational costs/year (Software + Hardware)		15% of software +hardware costs	31,380
Operational costs/year (System Administration and Maintenance)		20% of system development costs	122,000
Total Costs for Three Years (by using Net Present Value (NPV) approach and assuming a 6% discount rate)			**1,231,693**
Total Costs for Five Years (using NPV and by assuming 6% discount rate)			**1,691,832**

Table 7.1 includes the system development costs as well as the data warehousing operational costs. These costs were calculated by using a modified IT cost breakdown model originally presented by Love and Irani (2001), and the data warehousing cost breakdown suggested by Dyché (2000). It was assumed that an external consultant would be hired to develop the data warehouse while MDT's IT department would provide necessary logistic and organizational support and will participate in the process by giving feedback and user input. The data warehouse will be operational in 45-55 weeks (almost a year). The total costs for data warehousing implementation and operation are calculated for a period of 3 years and 5 years using the Net Present Value (NPV) approach and by assuming a discount rate of 6%. These costs are compared with the estimated savings expected to be achieved as a result of data warehousing implementation. This comparison is discussed in the next section.

7.3 Organizational Restructuring Model

Data warehouses have the potential to support significant changes in how an organization is structured and carries out its business. As pointed out in section 2.6.3, organizational design and the design of work processes are shaped by the amount and type of information required in a given environment and the organization's information processing capability (Galbraith, 1974; Tushman and Nadler, 1978). Because a data warehouse can provide more detailed, integrated, and historically complete information, it should be possible for an organization to operate very differently and "re-structure" itself (Watson et al., 2001).

Almost all research studies carried out to examine the effects of information systems implementation on organization design has been focused on manufacturing and service

159

industries. Although construction has the characteristics of both manufacturing and service industries yet it differs from these industries in various ways, due to its culture, work style, management principles and most importantly because of its unique products and processes. Hence it was necessary to conduct a focused research study to investigate the effects of information systems implementation on owner organizations in the construction industry. The *organizational restructuring model* developed as a part of the proposed decision-support framework in this study is an attempt to achieve this objective.

7.3.1 Experimental Setup for Developing Organizational Restructuring Model

The philosophy of action research approach is *action* and *evaluation*. As discussed in section 3.2.3, in this type of research the researcher intervenes a system through some action (i.e., by introducing a direct or an indirect change in the system) and then evaluates the effect of that action through various quantitative and qualitative measures. Data warehousing implementation was the *action* performed in this study. The next step was to *evaluate* its effects on organization design, and business and decision processes. In this section, the effects of data warehousing on organization design are discussed. The effects on business and decision processes are elaborated in sections 7.4 and 7.5.

Inline with the action research methodology, there are three ways to study the effects of different factors such as policy change, shift in organization strategies or effects of new information system implementation, etc. on organization design. The first option is to implement the changes and then study the actual organization for an experimental period of 6-24 months. The second option is to develop a dynamic simulation model of the organization (such as using System Dynamics concept) and investigate the effects of different factors (Ogunlana et al. 2003). The third option is to form a "Focus Group"

160

consisting of organization representatives and discuss all possible future scenarios using their experience and gut-feelings and then reach the optimum solution. All three techniques are well accepted in action research. The first option requires plenty of time while the second option needs good knowledge about systems simulation. The third option seemed most suitable choice due to the scope of this research study and limited availability of time. The focus group ensures full involvement of the organization executives and the results could lead to more realistic conclusions. Hence, it was decided to adopt this action research evaluation method to develop the organizational restructuring model.

A focus group may be defined as a group of individuals selected and assembled by researchers to discuss and comment on, from personal experience, the topic that is the subject of the research (Powell and Single, 1996). Focus groups are not polls but in-depth, qualitative interviews with a small number of carefully selected people who have full knowledge about the subject. The size of a focus group varies between six to twelve people. The focus group may work in different ways such as by conducting one or several moderated discussions of all participants, by continuously taking feedback of all participants and cross-examining them, by conducting a training and discussion workshop, or by using any combination of above-mentioned schemes (Gibbs, 1997).

The focus group for this research consisted of six representatives from the following functional areas of MDT.

(1) Contracts Administration

(2) Cost and Scheduling

(3) Construction Management Services

(4) Design and Engineering

(5) Computer Services

(6) Information Technology and Support Services

These personnel were selected based on their experience and continuous involvement in this research. They represented the four MDT divisions (i.e. Design & Engineering, Construction, Project Control and IT & Computer Services) which are involved in construction operations as well the IT division who is responsible for all IT/IS related operations. Hence it can be said that these personnel well represented the organization and had full knowledge about the organization activities and operations. It is important to note that some feedback was also collected from chiefs of three above-mentioned divisions. However, due to their busy and unpredictable time schedule; it was not possible to continuously involve them in discussions.

These personnel were also involved at earlier stages of this research such as owner organizational modeling, development of the data warehousing functional model and reference architecture and development of IT infrastructure model. The data warehousing model and the prototype system was shown to all these individuals to explain its functionalities and possible role in the future decision-support operations. After the demonstration, their feedback was collected on how the implementation of data warehousing would affect the existing organizational structure of MDT. Based on their input, several organizational redesign schemes were developed and discussed. After a series of three discussions, the focus group reached a final organizational restructuring model which is recommended for MDT as well as other similar construction owner organizations.

7.3.2. Assumptions

The organizational restructuring model is based on the following assumptions which were derived from previous experiences of data warehousing implementation in different manufacturing and services industries:

(1) A period of one year accompanied by suitable training is required to fully train executives about the use of data warehouse. Hence no organizational change should be implemented during this training and learning period (Rachmat, 2000).

(2) The post-implementation effects of data warehousing are slow, steady and continuous (Park, 2005). Therefore organizational restructuring should be conducted gradually preferably in two to three phases. There should be sufficient time gap between each phase so that the improvements in business and decision processes could be evaluated thoroughly before moving to the next phase.

(3) The data warehouse is used by the executives to make complex decisions quickly due to the availability of summarized, exceptional and detailed reports; data trends; and other decision-aids. Hence, the lower management levels that normally process such information and prepare reports for executives are no longer required (Subramanian, 1997).

(4) The initial restructuring efforts should be focused on the downsizing of departments or units where the human workload has been reduced after the implementation of data warehouse. In the later phase(s), several departments should be merged to form few new departments (Hwang, 2002).

7.3.3 Development of Organizational Restructuring Model

To develop the organizational restructuring model, the impact of data warehousing on different MDT divisions and management positions was examined. The purpose was to determine the reduction in work load and improvements in decision-making processes after the implementation of data warehousing system. The job functions of each management position before and after data warehousing implementation were discussed with the focus group to get their insight and feedback. Based on these discussions, two organizational restructuring models were proposed which are discussed in the following paragraphs. It is important to note that the currently proposed data warehouse is focused on construction operations and therefore no organizational change is proposed in the Design and Engineering division.

Inline with assumptions 1-4, it was decided that the organizational restructuring efforts would start after one year of data warehousing implementation and carried out in two phases. In the first phase, the Construction division and Project Control division would be downsized while in the second phase, they would be merged to form a new division. These organizational changes would help to decrease the hierarchy in the organizational structure and result in substantial cost savings.

The existing organizational structure of Construction and Project Control divisions is shown in Figure 7.6. Currently, there are approximately 40 employees who are working in these two divisions. During the group discussions, it was determined that most of the computational work is performed by the section managers. The section managers collect project data from the office and field staff; organize and analyze the data; prepare reports; and send these reports to respective mangers or division chiefs for necessary decision-

164

making. After implementation, these roles would be carried out by the data warehousing system and the services of section managers may no longer be required. Hence, it was decided to eliminate these positions after one year of data warehousing implementation. The new organizational structure (hereafter called organizational structure no. 2) is illustrated in Figure 7.7. The peculiar features of this organizational structure are reduced organizational hierarchy and downsizing of management staff by 23%. The effect of downsizing on organizational direct costs is further discussed in section 7.3.4.

It was further decided that this new organizational structure should persist for a period of 2 years (3 years in total after including 1 year of training). During this time, the division chiefs and managers would have developed necessary expertise to use data warehousing system for their daily, short-term and long-term business needs. At the same time, the data warehouse would have been populated with complete historic and current project data. Only after meeting these milestones, the second phase of organizational restructuring should be carried out.

During phase 1, it was determined that different functional areas of project management have been split between the two divisions and as a result, unnecessary time and efforts are wasted in coordination and decision-making. For example, the work measurement (and inspections) is performed by Construction division while cost control (and project monitoring) is carried out by Project control division. As both phases often go side by side during a project, the two divisions have to consult each other and wait for necessary approvals and decision-making. This process could be made more effective if both functional areas are merged to work under single management. Based on this concept, the second phase of organizational restructuring was carried out as shown in

165

Figure 7.8 (hereafter called organizational structure no. 3). In this phase, it is proposed to merge Construction division and Project Control division to form a new "Construction Planning and Management division". This new division would carry out all operations related with construction planning, management, and control. As the division would work under a single division chief, the decision-making will be more quick and effective. It is further expected that this change would increase coordination between the employees and results in improved productivity. The number of employees would now reduce to 26 thereby indicating a 35% reduction in workforce as compared to the existing one.

In several public owner organizations, the decision of divisions' merger could be very difficult due to different governmental, political, social or internal organizational limitations. It is suggested that such organizations should carry out at least phase 1 of organizational structuring to ascertain certain direct cost savings in a long run. The effect of both phases of organizational restructuring on direct cost savings for a period of 3, 5 and 10 years is discussed in the following section. The opinions of a selected group of 20 owner organizations about both phases of restructuring are shown in section 7.5.2.

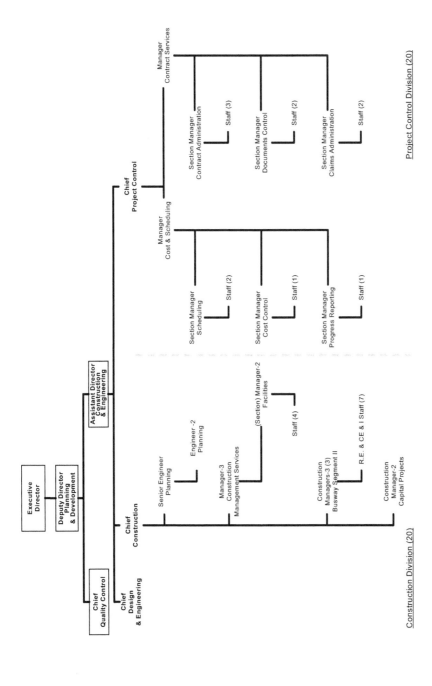

Figure 7.6: Existing Organizational Structure of MDT

167

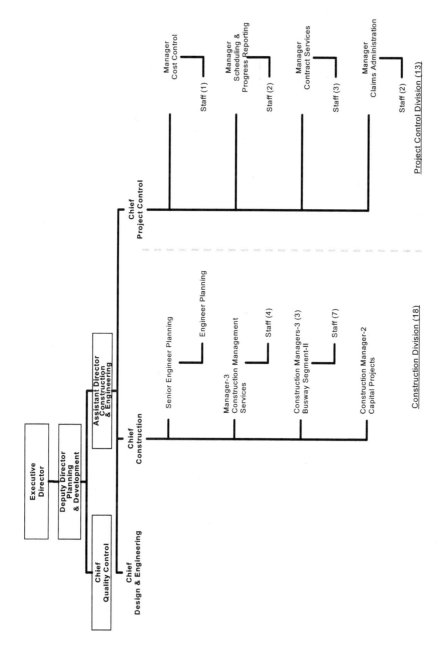

Figure 7.7: Proposed Organizational Structure after One Year of Data Warehousing Implementation

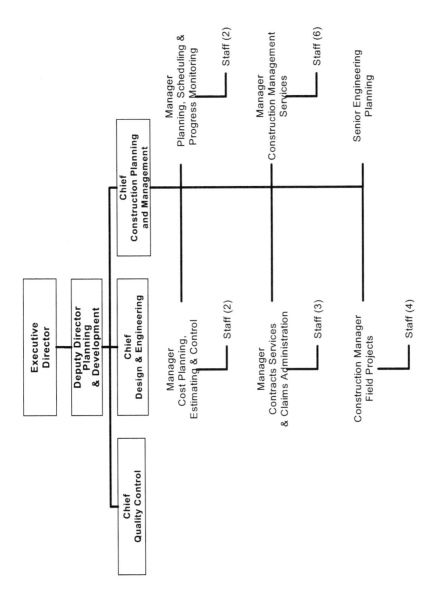

Figure 7.8: Proposed Organizational Structure after three Years of Data Warehousing Implementation

7.3.4 Comparison of Direct Cost Savings

Table 7.2 shows the estimated direct cost savings at different periods of time which would be realized after data warehousing implementation. The Net Present Value (NPV) method is used to convert all future cost savings into present day dollars.

Table 7.2: Estimated Cost Savings after Organizational Restructuring (Direct Costs Only)

Management Positions Added/Removed	No.	Unit Savings* ($)	Total Savings ($)
Organizational Structure # 2 (after one year of data warehousing implementation)			
Section Managers	-10	45,000	450,000
Staff	-2	35,000	70,000
New Manager Positions	+3	65,000	(195,000)
Net savings per year			325,000
Net savings after 3 years of data warehousing implementation (using NPV and assuming 6% discount rate)			*795,917*
Organizational Structure # 3 (after three years of data warehousing implementation)			
Associate Director	-1	128,000	128,000
Division Chief	-1	102,000	105,000
Managers	-2	65,000	130,000
Staff	-4	35,000	140,000
Total savings per year			503,000
Net savings after 5 years of data warehousing implementation (using NPV and assuming 6% discount rate)			*3,092,461*

* The unit savings represent the average annual employees' salary. These salary figures are taken from Job Description/Pay Plans Document published by the MDT Employee Relations Department (2005).

Table 7.3 compares the costs incurred for data warehousing implementation (shown in Table 7.1) with the resulting direct cost savings which would be achieved as a result of changes in the organizational structure.

170

Table 7.3: Comparison of Data Warehousing Implementation Cost and Net Associated Direct Cost Savings

Time after Data Warehousing Implementation	Initial and Operational Costs ($)	Savings due to Organizational Restructuring ($)	Net Savings ($)
Organizational structure # 2			
3 years	1,231,639	795,917	-435,722
5 years	1,691,832	1,835,485	143,653
10 years	2,465,429	6,047,705	3,582,276
Organizational structure # 3			
5 years	1,691,832	3,092,461	1,400,629
10 years	2,465,429	11,090,792	8,625,363

With the organizational structure no. 2, a period of around five years would be required to payoff the data warehousing costs by means of savings in direct salary costs. After five years, MDT would start realizing substantial reductions in direct salary costs. At the end of 10 years, it is expected that the net savings due to direct salary costs would be approximately $3.5 million.

If the organizational structure no. 3 would have been incorporated after 3 years of data warehousing implementation, net savings in direct salary costs will be $1.4 million after 5 years and $8.6 million after 10 years. These figures indicate that the merger of both divisions could increase the net savings by $5.1 million which is approximately 2.5 times more than the savings realized with organizational structure no. 2 only.

It is important to note that the data warehousing will also result in substantial savings in indirect costs such as information management and processing costs, report preparation costs and other miscellaneous operational costs. However, these costs are not included here because the focus of this comparison is to show the benefits of organizational restructuring on direct costs savings.

7.4 The Decision-Support Framework

The final decision-support framework is shown in Figure 7.7. The framework presents a five phase scheme to plan, design, build and implement data warehouses for the public construction owner organizations. Each stage is divided into several tasks which are explained along with suggested tools to accomplish these tasks. This framework can be used as a road map during the life cycle of a data warehouse. Since the nature and sequence of construction operations in most public owner organizations is same, and the decision-making requirements are not much different, the functional and implementation models developed at different stages of this research could be used for other organizations with little modifications. This effort will save tremendous amount of time for these organizations and even their IT departments (those who have no prior experience of data warehousing) may initiate such projects with no or limited help from the outside consultants. This will result in substantial cost savings. This research also provides an adequate idea about data warehousing implementation costs and resulting savings in direct and operational costs. The executives can use these cost comparisons for reaching a decision about data warehousing implementation in their organization. The organizations may form a focus group and consider a hypothetical implementation of the data warehouse in their organization to evaluate its impact on organization design, productivity, decision-support capabilities and other organization-specific factors. This practice will help to quantify the data warehousing implementation benefits before proceeding towards the actual design.

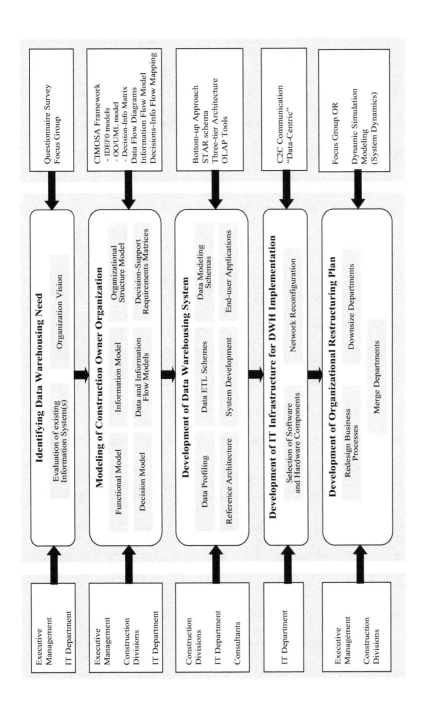

Figure 7.9: The Proposed Decision-Support Framework

173

7.5 Validation of the Framework

The decision-support framework was validated using several quantitative and qualitative measures to ensure that the research objectives have been achieved. The validation procedure was conducted in two stages. In the first stage, the framework was validated by the Miami-Dade Transit (MDT) while in the second stage; it was validated by other construction owner organizations. The questionnaire survey approach was adopted for validation of the framework at both stages. The structure of the questionnaire and the statistical analysis of the data obtained through the surveys are explained in the following sections.

7.5.1 Validation by the Miami-Dade Transit (MDT)

The questionnaire survey no. 2, as shown in Appendix D, was used to validate the framework by the MDT. The questionnaire was distributed among the six members of the focus group which were mentioned in section 7.3.1. These respondents represented different participating divisions of MDT and were well familiar with every stage of the framework development and implementation. Hence their responses could be considered as unbiased and fairly realistic.

The first question in the survey illustrates various qualitative measures which could used to evaluate the performance of the decision-support framework. These qualitative measures are proposed by different researchers as a way to gage the performance of a data warehouse (Park, 2005; Dyché, 2000; Gray and Watson, 1998). Part (a) of this question is focused on the improvements in the data management operations while part (b) concentrates on the improvements in the decision-support operations as a result of the framework implementation. Tables 7.3 and 7.4 illustrate the opinions of the respondents.

174

Due to the small size of the sample, respondent's response was evaluated in numbers instead of percentage.

Table 7.4: Improvements in the Data Management Operations at MDT

Satisfaction Criteria	Mean	Standard Deviation	Level of Improvement				
			No (1)	Little (2)	Some (3)	Fair (4)	Significant (5)
Ease in data access	4.33	0.82	-	-	1	2	3
Data quality	4.67	0.82	-	-	-	2	4
Data integration	4.50	0.52	-	-	-	3	3
Response time	4.17	0.98	-	-	2	1	3
Data trends/ summaries	4.83	0.41	-	-	-	1	5
Quality of reports	3.67	1.21	-	1	2	1	2

Table 7.5: Improvements in the Decision-Support Operations at MDT

Satisfaction Criteria	Mean	Standard Deviation	Level of Improvement				
			No (1)	Little (2)	Some (3)	Fair (4)	Significant (5)
Every day decisions	4.67	0.52	-	-	-	2	4
Short and long term planning	4.83	0.41	-	-	-	1	5
Confidence in decisions	4.17	0.98	-	-	2	1	3
Time to make decisions	4.33	0.82	-	-	1	2	3
Personal productivity	4.67	0.89	-	-	-	2	4
Organizational productivity	4.00	1.21	-	-	2	2	2

The results summarized in both tables indicate that the Mean of responses in all cases except one is four or higher. This shows that the respondents felt that there would be fair-to-significant improvements in all processes as a result of framework implementation. Some of these satisfaction criteria were also used during the first questionnaire survey to measure the respondent's level of satisfaction with their existing information systems. The comparison of results obtained from both surveys could be used as an indicator to gage any improvements in the process. For this purpose, the Kruskal-Wallis test was performed to compare the Mean of responses obtained through both questionnaire surveys. The results are shown in Table 7.6.

Table 7.6: Comparison in Performance before and after Implementation of Proposed Decision-Support Framework

Comparison Criteria	Mean of Responses		Sum of Squares (X^2)	Significant error (p)
	Without framework	With framework		
Ease in data access	2.83	4.33	8.42	0.01
Data quality	3.12	4.67	6.97	0.03
Data integration	2.45	4.56	7.01	0.02
Productivity improvement	1.98	4.67	9.14	0.01
Quality of reports	2.78	3.67	5.12	0.03
Support for every day decisions	2.46	4.67	10.12	0.00
Support for short and long term planning	1.78	4.83	6.98	0.01

The results indicate that the significant error (p) in all cases is below 0.05 (<5%). This means that the results are statistically significant, i.e. a significant difference exists between the earlier and later responses of the same respondents. This analysis could be used to further strengthen the claims that the proposed framework would induce significant improvements in data management and decision-support processes.

In the question number 2, the respondents were asked to indicate potentials time savings (e.g. in decision-making, reports generation etc.) and operational cost savings (e.g. costs associated with information gathering, reports generation, etc.) which would be obtained after implementing the proposed decision-support framework. Their responses are summarized in Table 7.7. The Mean of responses for time savings indicates that approximately 20%-30% time could be saved in various operations after the framework implementation. The corresponding operational cost savings are between 10% to 20%. These cost savings are in addition to direct cost savings which would be achieved after the restructuring of the organization.

Table 7.7: Potential Cost and Time Savings obtained after the Framework

Implementation

Factor	Mean	Standard Deviation	Amount of Savings (%)					
			< 10% (1)	10-20% (2)	20-30% (3)	30-40% (4)	40-50% (5)	>50% (6)
Time savings	2.67	0.82	-	3	2	1	-	-
Operational cost savings	2.17	0.75	1	3	2	-	-	-

The results of both questions clearly indicate that the proposed decision-support framework would lead to significant improvements in the data management and decision-support practices at MDT and also results in substantial cost savings.

7.5.2 Validation by Other Construction Owner Organizations

Twenty public construction owner organizations representing various categories as specified in Table 4.1 were chosen to validate the framework. These organizations were chosen from a list of preselected companies which satisfied two criteria, (1) they showed a need of data warehousing implementation in their organization during the questionnaire 1, and (2) they use their existing information system(s) for short and long-term decision-making but not satisfied with its performance. Simple random sampling was performed to select a sample using 90% confidence level and 5% confidence interval. Table 7.8 illustrates the list of selected public construction owner organizations which participated in the validation process.

177

Table 7.8: Public Construction Owner Organizations Selected for Framework Validation*

xxxxxx Department of Transportation	Port of xxxxxx
xxxxxx Department of Transportation	xxxxxx Transit Authority
xxxxxx Department of Transportation	xxxxxx Transit Authority
xxxxxx Department of Transportation	xxxxxx Rapid Transit Authority
xxxxxx Department of Transportation	xxxxxx Railroad Company
xxxxxx Department of Transportation	xxxxxx Gas and Electric Company
xxxxxx Port Administration	xxxxxx General Electric, xxxx
Port of xxxxxx, xxxxxx	xxxxxx Metropolitan Area Transit Authority
xxxxxx Port Authority	xxxxxx Electric Company, xxxx
Port Authority of xxxxxx & xxxxxx	xxxxxx County Metropolitan Transportation Authority

* Due to the data privacy issues, the full names of the organizations are not shown.

Before sending the questionnaire survey, the respondents (located from questionnaire 1) in these organizations were contacted. They were briefly explained about the nature of the research, potential findings and their role as a respondent to validate the framework. After this arrangement, the questionnaire survey number 3, as shown in Appendix E, and a brief document illustrating the different stages of the decision-support framework development and implementation for MDT were sent to them. A week after dispatching of the surveys, all respondents were again contacted to answer any queries or concerns which they might have about the framework. It was ensured that all the respondents had correctly understood the framework and the questionnaire survey. This step was necessary to remove any bias in results which might develop due to different interpretations of the proposed decision-support framework by the respondents. All the respondents answered the questionnaire survey thereby representing a 100% response rate.

In the first question, the respondents' opinion was sought about the possibility of proposed decision-support framework implementation in their organization. Fifteen respondents (75%) answered Yes, four (20%) were undecided while only one (5%)

answered No. The respondents who were undecided or answered No were further asked about the main reasons behind their decision. Their answers in order of priority were as follows: (1) Senior management may prefer a different solution; (2) Organization is in a process of implementing a new information system, and (3) Lack of financial resources. These responses indicated that the proposed framework may still be technically feasible but not needed at this stage due to one or other reasons.

The question no. 2 was designed to gage improvements in different data management and decision-making practices as a result of the decision-support framework implementation. The results are summarized in Tables 7.9 and 7.10.

Table 7.9: Improvements in the Data Management Operations

Satisfaction Criteria	Mean	Standard Deviation	Level of Improvement				
			No (1)	Little (2)	Some (3)	Fair (4)	Significant (5)
Ease in data access	3.70	1.06	-	3 (15%)	5 (25%)	7 (35%)	5 (25%)
Data quality	4.15	0.87	-	1 (5%)	3 (15%)	8 (40%)	8 (40%)
Data integration	4.05	1.06	-	2 (10%)	4 (20%)	5 (25%)	9 (45%)
Response time	4.40	0.91	-	1 (5%)	2 (10%)	5 (25%)	12 (60%)
Data trends/summaries	4.00	1.02	-	2 (10%)	4 (20%)	6 (30%)	8 (40%)
Quality of reports	3.75	0.98	-	3 (15%)	4 (20%)	8 (40%)	5 (25%)

Table 7.10: Improvements in the Decision-Making Operations

Satisfaction Criteria	Mean	Standard Deviation	Level of Improvement				
			No (1)	Little (2)	Some (3)	Fair (4)	Significant (5)
Every day decisions	4.15	0.77	-	-	5 (20%)	7 (35%)	8 (40%)
Short and long term planning	4.30	0.73	-	-	3 (15%)	8 (40%)	9 (45%)
Confidence in decisions	4.20	0.90	-	1 (5%)	3 (15%)	7 (35%)	9 (45%)
Time to make decisions	3.85	0.89	-	2 (10%)	4 (20%)	9 (45%)	5 (25%)
Personal productivity	4.00	0.84	-	-	6 (30%)	8 (40%)	6 (30%)
Organizational productivity	3.90	0.81	-	-	8 (40%)	6 (30%)	6 (30%)

The Mean of responses vary between 3.5 to 4.5, which indicates that most of the respondents thought that the decision-support framework would result in fair-to-significant improvements in different processes. It also indicates that the opinions of respondents from other construction owner organizations were not different from the opinions shown by respondents at MDT.

Similar to section 7.5.1., a Kruskal-Wallis test was performed to statically measure the differences in the opinions of the same respondents before and after the implementation of the decision-support framework. The results are shown in Table 7.11. The significance error for all performance measures came out to be below 0.05 which indicates that the results are statistically significant.

Table 7.11: Comparison in Performance before and After Implementation of Proposed

Decision-Support Framework

Comparison Criteria	Mean of Responses		Sum of Squares (X^2)	Significant error (p)
	Without framework	With framework		
Ease in data access	3.01	3.70	5.41	0.02
Data quality	3.36	4.15	6.39	0.02
Data integration	3.54	4.05	4.96	0.04
Productivity improvement	1.76	4.00	8.95	0.01
Quality of reports	2.98	3.75	4.25	0.03
Support for every day decisions	1.49	4.15	10.58	0.00
Support for short and long term planning	2.11	4.30	7.25	0.00

In question no. 3, the respondents were asked about potentials time savings and operational cost savings which might be obtained after implementing the proposed decision-support framework. Their responses are summarized in Table 7.12. The Mean of responses for time savings indicates that approximately 20%-30% time could be saved in various operations after the framework implementation. The resulting operational cost savings are up to 20%. Again, the trend is very similar to the MDT respondents.

Table 7.12: Cost and Time Savings after using the Proposed Framework

Factor	Mean	Standard Deviation	Amount of Savings (%)					
			< 10% (1)	10-20% (2)	20-30% (3)	30-40% (4)	40-50% (5)	>50% (6)
Time savings	2.70	1.03	3 (15%)	5 (25%)	7 (35%)	5 (25%)	-	-
Operational cost savings	2.00	0.65	4 (20%)	12 (60%)	4 (20%)	-	-	-

The question no. 4 was about the organizational restructuring after the data warehousing implementation. Thirteen (65%) respondents favored the organizational redesign while four respondents (20%) were undecided and three (15%) answered No. Among the fourteen respondents who showed positive response, four (28%) favored the idea of entire organizational restructuring, seven (50%) backed the idea of restructuring of individual departments while three (21%) thought that there should be some changes in each department without going though any major restructuring effort. These responses indicate that most of the respondents agreed that the organizational restructuring is essential to reap the full benefits of the data warehousing implementation however the amount of restructuring effort varies from organization to organization depending on the organization goals, vision and existing structure.

In the last question, the respondents were asked if they foresee an actual implementation of this or a similar decision-support framework in their organizations. Eight (40%) respondents said Yes, seven (35%) were undecided while five (25%) answered No. Among the eight respondents who answered Yes, two mentioned a possible implementation within the next 3 years, five indicated within the next 5 years while one respondent thought about the next 10 years.

The results of this questionnaire survey strongly suggest that the proposed decision-support framework is valid for most of the public construction owner organizations. The various functional and implementation models which were originally developed for MDT are equally valid for other public owner organizations with minor modifications. This means that the other organizations do not need to "reinvent the wheel" and they could use these models as a road-map to plan and build their data warehouses.

7.6 Summary

The proposed decision-support framework consists of five phases namely data warehousing needs identification, organizational modeling, development of functional model and reference architecture of the data warehouse, IT infrastructure model and the organizational restructuring model. It was found that the successful implementation of a data warehouse requires substantial changes in the IT infrastructure and the organizational structure. The organization redesigning could result in substantial savings in direct salary costs. Further 10-20% savings could be achieved in operational and other indirect costs. The proposed decision-support framework was validated by MDT as well as twenty different public construction owner organizations. The results indicate that significant improvements would be realized in data management and decision-making practices after the implementation of the proposed decision-support framework.

CHAPTER 8

CONCLUDING REMARKS

8.1 Research Summary

The project data in construction owner organizations are typically stored in Online
Transaction Processing (OLTP) systems to support day-to-day construction operations.
These data are often non-validated, non-integrated and stored in forms and formats which
makes it difficult for decision-makers to make timely decisions. Due to these limitations
of OLTP systems, decision-makers at many occasions have to use their experience and
gut-feelings to make the critical project decisions. It is also recognized that information
systems in construction owner organizations are usually not compatible with their
existing IT infrastructure and organizational structure. This lack of compatibility further
reduces the effectiveness of information systems; and results in higher operational costs
and lower productivity.

The purpose of this research was to develop a framework to guide storage and
retrieval of validated and integrated data for timely decision-making and to enable
construction owner organizations to redesign their organizational structure and IT
infrastructure matched with information system capabilities. Due to inherent limitations
of OLTP systems, *data warehousing* a fairly recent database management technique was
utilized. Data warehousing is an improved approach for integrating data from multiple,
often very large, distributed, heterogeneous databases and other information sources. It is
based on Online Analytical Processing (OLAP) concept as opposed to OLTP. The OLAP
facilitates multidimensional analysis of data using various statistical and analytical
techniques, and effectively supports planning and decision-making processes.

A relatively new philosophy of research named *action research* was employed in this study. Action research is an iterative technique in which the researcher investigates the problem domain, identifies the problem, gets involved in introducing some changes to improve the situation and evaluates the effects of those changes. It is a novel approach for building and testing theory within context of solving an immediate practical problem in a real setting.

The research was carried out in four distinct but interrelated phases. First, a questionnaire survey was conducted among the construction owner organizations. The purpose was to assess the data management practices employed in these organizations and degree of utilization of existing information systems (IS) in planning and decision-making. It was found that medium to large size construction owner organizations use a formal database management system to store and manage project data while small size organizations rely on the traditional paper and folder techniques. The main utilization of project data was found in preparing summarized reports for project monitoring and control. Hardly few organizations were found to effectively utilize their project data for formal planning and decision-making operations. The results of this questionnaire survey also validated the initial research hypothesis that construction owner organizations do not effectively utilize their project data for planning and decision-making due to lack of decision-support in their existing information systems (IS).

In the second phase, inline with action research methodology, one construction owner organization Miami-Dade transit (MDT) was chosen as a research collaborator. MDT is in charge of administration of all public transportation-related construction projects in the Miami-Dade County of the state of Florida. In the first step, the multi-perspective

185

modeling of MDT was carried out to understand its existing organizational structure, business functions and processes, data management practices and decision-making procedures. In the next step, data and information flow models were prepared to identify the flow of operational data into, within and outside the organization. In the last step, decision-support requirements matrices were prepared by mapping data and information flow models with the multi-perspective models developed in step 1. These matrices were used to identify the user requirements for the development of the data warehousing system. The modeling efforts carried out in this phase provides a systematic enterprise modeling approach for construction owner organizations.

In the third phase, the functional model and reference architecture of the data warehouse were prepared using the *bottom-up* approach and *3-tier* architecture. The functional model illustrates how the system would work while the reference architecture reveals its physical and technical implementation using different software and hardware tools. The data modeling was performed using the *Star* schema. A prototype data warehousing system was developed and verified. It is important to note that although different models in phase 2 and 3 were particularly developed for MDT but they are generic to most construction owner organizations and can be used with some modifications.

During the course of this research, it was recognized that successful implementation of data warehousing requires appropriate changes in the organizational structure and IT infrastructure. As a result, the revised IT infrastructure model and the organizational restructuring model were proposed in the last phase of this research. The IT infrastructure model was developed using the *data-centric* and *center-to-center communication* approaches. This new model ensured that the project data from different project sites and stakeholders were securely transferred to the data warehouse and all users would have the same version of the data at any given time. The organizational restructuring model portrayed two phases to redesign the organization after implementation of data warehousing system. The first phase would eliminate all lower management positions which are typically involved in data analysis and reports preparation. The reason is that the data warehouse would perform all these tasks and hence these positions might no longer be required. The second phase would result in merger of two functional departments to form a single new department. The bases are to bring all split functional areas under one management and save time and efforts which are otherwise wasted in coordination and distributed decision-making. Cost analysis showed that the organization redesigning could result in substantial savings in direct salary costs.

The end product of this research is a decision-support framework which provides schematic procedures to design and implement the data warehousing technology for public construction owner organizations. This framework consists of five phases namely data warehousing needs identification, organizational modeling, development of functional model and reference architecture of the data warehouse, IT infrastructure model and the organizational restructuring model. The framework was validated by MDT as well as twenty different public construction owner organizations. The results indicated significant improvements in data management and decision-making practices, and substantial time and operational cost savings.

8.2 Conclusions

The following concluding remarks can be made on the basis of the findings of this research:

This research study was based on a hypothesis that the construction owner organizations do not effectively utilize project data for planning and decision-making. The results of the first questionnaire survey validated this hypothesis. The results indicated that only 40% of surveyed organizations (65 out of 163) employ a formal database management system to store and manage the project data. It was also found that 39% percent of these 65 organizations (or 25 organizations) use their existing information systems to make everyday decisions while only 29% (or 19 organizations) use them for short and long term planning. The reasons found were low decision-support capabilities of existing information systems, poor data quality, difficult data access and high cost of operations. Ninety-one percent surveyed organizations indicated that they require a new and enhanced information system to support planning and decision-making operations.

It was demonstrated in this research that data warehousing could be used as an effective technique to solve the current decision-support problems in the public construction owner organizations. Fair-to-significant levels of improvements were recorded in data management and decision-support operations after the implementation of data warehousing system. These improvements were examined using both qualitative (e.g. data access, data quality, response time, personal and organizational productivity, etc.) and quantitative measures (e.g. operational time and cost savings). Approximately 20%-30% operational time savings and 10%-20% operational cost savings were realized through use of data warehousing system.

The results indicated that replacing an existing information system with data warehouse led to limited benefits only such as better decision-support, improved data quality and easy data access. The full benefits of data warehousing were realized when subsequent changes were made in the existing organizational structure and IT infrastructure.

It was established that the restructuring of public construction owner organizations should be carried out in two phases after one year of data warehousing implementation. The one year time would ensure sufficient users' training about data warehousing use. In the first phase, individual departments should be downsized by eliminating lower management positions whose work tasks would be performed by the data warehouse. In the second phase, merger of two (or more) departments which have overlapping construction-related operations should be carried out.

A common problem found in most public construction owner organizations was that different stakeholders have different versions of the same data at a given time because the data storage and updating operations were asynchronous. The proposed IT infrastructure model provides a synchronous mechanism to collect and share project data in real time through data warehousing. This would lead to reduced disputes and claims which arise as a result of miscommunication and lack of coordination.

The research demonstrated that although no formal methods are available for modeling construction owner organizations, the *enterprise modeling* techniques originally developed for manufacturing and service industries could be successfully used with some modifications. The research presented a framework consisted of modified CIMOSA reference architecture to model the organizational, functional, informational and decision-making perspectives of construction owner organizations. These models were used to ascertain user requirements for developing the data warehouse.

8.3 Research Contributions

This research study contributed to the existing body of knowledge in many ways, the most significant ones are listed below.

It provided a feasible and adoptable solution to solve the existing data quality and decision-support problems in the construction owner organizations. The literature review suggested that these problems have been realized by different researchers since several decades but no satisfactory solution was developed. Construction is one of the industries which always fall behind in the adoption of new automation techniques and novel computational approaches due to its traditional structure, norms and values. Although the data warehousing technique is known from a decade, it was hardly employed in the construction industry. The few earlier research efforts as mentioned in the literature review were focused on the development of broad-scope conceptual models only. This is the first study where concept of data warehousing has actually been tested in a construction organization and its post-implementation effects are evaluated. It is expected that the decision-support framework proposed in this research would serve as a roadmap for other construction organizations to plan, design, build and implement data warehouses.

This research made it clear that successful data warehousing implementation in construction owner organizations would require restructuring of the organizational setup. Although this fact was pointed out by some researchers but no planned research effort was found in the literature. Besides, the author is not aware of any existing research in construction where the effects of information systems implementation were investigated on the organizational structure. The result of this study enabled to correlate the effects of information systems implementation in general, and data warehousing in particular, with the organizational structure. These results are not only beneficial for the construction industry, they also provide a lead to organizational restructuring in other industries.

8.4 Suggestions for Future Research

The research findings and conclusions are limited by the scope of this research. An outline of suggestions for carrying out future research on this subject is give below:

The enterprise modeling is required to understand the operations of a construction organization and to capture user requirements for data warehousing development. As no construction-specific enterprise modeling techniques are available, there is a need to modify the existing techniques developed for manufacturing and service industries, and to make them compatible with the requirements of the construction industry. Such research would lead to development of an enterprise modeling framework for different types of construction organizations such as owners, contractors, consultants, etc.

The current research was focused on public construction owner organizations. A separate research should be conducted to develop data warehouses for private construction owner organizations, consultants, and contractors. This research would help to compare the effects of data warehousing implementation on different types of construction organizations and assist in selecting the best candidate organizations.

A study should be conducted to investigate the *pros and cons* of data sharing through data warehouse between different construction stakeholders such as owners, contractors and consultants. The study should be focused on the data ownership problems, level of

191

data access to different stakeholders, and data security and vandalism issues. These issues were raised by different construction owner organizations involved in this study.

A data warehouse supports two types of decision-support applications, OLAP-based applications and data mining applications. The OLAP-based applications were examined in this research. A separate study should be conducted to investigate the suitability of different data mining applications for construction organizations. Data mining could be extremely helpful in planning operations to identify data trends, data associations and development of forecasting models using neural networks.

In this research, the organizational restructuring model was developed using the opinions of a focus group. Another way to achieve the same objective is through system simulation. Simulation provides a more scientific approach to test the effects of various parameters on organizational design. Among various simulation techniques available, it is suggested to use *system dynamics approach*. System dynamics approach is based on cause-effect relationships to examine the effects of various parameters on the subject of interest (e.g. organizational design) over a period of time. This method would enable developing a dynamic organizational model to test several hypothetical organizational designs before arriving at the optimum design.

REFERENCES

Abudayyeh, O., Temel, B., Al-Tabatabai, H., and Hurley, B. (2001). "An Intranet-based Cost Control System." *Advances in Engineering Software*, 32 (2), 87-94.

Ahmad, I. (1990). "Decision-Support System for Modeling Bid/No-Bid Decision Problem." *Journal of Construction Engineering and Management*, 116(4), 595-608.

Ahmad, I. (2000). "Data Warehousing in Construction Organizations." *Proceedings of the ASCE 6th Construction Congress*, Orlando, Florida, 20-26.

Ahmad, I., and Azhar, S. (2002). "Data Warehousing in Construction: From Conception to Application." *Proceedings of the First International Conference on Construction in the 21st Century: Challenges and Opportunities in Management and Technology*, April 24-26, 2002, Miami, Florida, USA, 739-747.

Ahmad, I., and Azhar, S. (2005). "Implementing Data Warehousing in the Construction Industry: Opportunities and Challenges." *Proceedings of the Third International Conference on Construction in the 21st Century (CITC-III)*, Athens, Greece, September 15-17, 863-871.

Ahmad, I., and Nunoo, C. (1999). "Data warehousing in the Construction Industry: Organizing and Processing Data for Decision-Making." *Proceedings of the 8th International Conference on Durability of Building Material and Components*, Institute for Research in Construction, Vancouver, British Columbia, 1-10.

Ahmad, I., Azhar, S., and Lukauskis, P. (2004). "Development of a Decision Support System using Data Warehousing to Assist Builders/Developers in Site Selection." *Automation in Construction,* 13(4), 525-542.

Ahmad. I., and Sein, M. K. (1997). "Building Construction Project Teams for TQM: A Factor-Element Impact Model." *Journal of Construction Management and Economics*, 15(5), 457-467.

Ahmed, S.M., Ahmad, I., Azhar, S., Mallikarjuna, S., and Kappaguntula, P. (2003). "Implementation of Enterprise Resource Planning (ERP) Systems in the Construction Industry: A Study Extended." *Proceedings of the Third International Conference on Information Systems in Engineering and Construction* (on CD Rom), Cocoa Beach, Florida, June 12-13.

Al-Deek, H. (2004). *The Central Florida Data Warehouse, Phase-2*, Final Report, University of Central Florida, Orlando and Florida Department of Transportation, June 2004.

Al-Deek, H., and Abd-Elrahman, A. (2002). *An Evaluation Plan for the Conceptual Design of the Florida Transportation Data Warehouse, Phase-1*, University of Central Florida, Orlando and Florida Department of Transportation, March 2002.

Akintoye, A.S., and Macleod, M.J. (1997). "Risk Analysis and Management in the Construction." *International Journal of Project Management*, 15(1), 31-38.

Amor, R., and Anumba, C.J. (1999). "A Survey and Analysis of Integrated Project Databases." *Proceedings of the International Conference on Concurrent Engineering in Construction: Challenges for the New Millennium*, CIB Publication 236, Espoo, Finland, August 25-27, 217-227.

Ang, J., and Teo, T.S.H. (2000). "Management Issues in Data Warehousing: Insight from the Housing and Development Board." *Decision Support Systems*, 29, 11-20.

Anumba, C.J. (2000). "Integrated Systems for Construction: Challenges for the Millennium." *Proceedings of the First International Conference on Implementing IT to Obtain Competitive Advantage in the 21st Century*, January 17-19, Hong Kong, 78-92.

Aouad, G., Kirkham, J., Brandon, P., Brown, F., Child, T., Cooper, G., Ford, S., Oxman, R., and Young, B. (1995). "The Conceptual Modeling of Construction Management Information." *Automation in Construction*, 3, 267-282.

Argyris, C. and Schön, D. (1978) *Organizational Learning: A Theory of Action Perspective*, Addison-Wesley, Reading, MA.

Azhar, S. (2007). "Improving Collaboration between Researchers and Practitioners in Construction Research Projects using Action Research Technique." Proceedings of the 43rd ASC National Annual Conference, Flagstaff, AZ, April 12-14 (on CD ROM).

Ballard, C., Herreman, D., Schau, D., and Bell, Rhonda. (1998). *Data Modeling Techniques for Data Warehousing*, International Technical Support Organization, IBM, MI.

Barker, K. (2000). *Data Warehousing Overview*, University of Calgary, Alberta, Canada.

Barnes, T. (1998). *Statistical Methods for Engineers*, John Willey, NY.

Baskerville, R. (1997). "Distinguishing Action Research from Participative Case Studies." *Journal of Systems and Information Technology*, 1(1), 25-45.

Baskerville, R. (1999). "Investigating Information Systems with Action Research." *Communications of the Association of Information Systems*, 2(19), 7-17.

194

Baskerville, R.L., and Wood-Harper, A.T. (1996). "A Critical Perspective on Action Research as a Method for Information Systems Research." *Journal of Information Technology*, 11, 235-246.

Bjork, B.C. (1999). "Information Technology in Construction: Domain Definition and Research Issues." *International Journal of Computer-Integrated Design and Construction*, 1(1), 3-16.

Booch, G., Rumbaugh, J., Jacobson, I. (1999). *Unified Modeling Language User Guide*, Addison-Wesley, Reading, MA.

Bouma, G., and Atkinson, G. (155). *A Handbook of Social Science Research: A Comprehensive and Practical Guide for Students*, Oxford University Press, UK.

BRE. (2002). "BRE Department Research Activities." Online at http://www.bre.polyu.edu.hk. Accessed on November 12, 2002.

Chahine, J.R., and Janson, B.R. (1987). "Interfacing Databases with Expert Systems: A Retaining Wall Management Application." *Microcomputer in Civil Engineering*, 2(1), 11-17.

Chau, K.W.; Cao, Y; Anson, M; and Zhang, J. (2003). "Application of Data Warehouse and Decision Support System in Construction Management." *Automation in Construction*, 12(2), 213-224.

Chaudhuri, S. and Dayal, U. (1997)."OLAP Technology and Data Warehousing." *ACM SIGMOD RECORDS*, Heward Packard Labs.

Chen, C., Udo-Inyang, P., and Schmitt, F. (1994). "Integration of a Database Management System and a Knowledge-Based Expert System in Construction: A Review." *On-line at* http://asceditor.unl.edu/archives/1994/chen94.htm, Accessed on November 13, 2002.

Chen, P.H., Wan, C., Tiong, R.L.K., Ting, S.K. and Yang, Q. (2004). "Augmented IDEF1-based Process-Oriented Information Modeling." *Automation in Construction*, 13(6), 735-750.

CIFE. (2002). "Introduction to 4D Research at Stanford, Dept. of Civil and Environmental Engineering and the Center for Integrated Facility Engineering." Online at http://www.stanford.edu/group/4D/4D-home.htm. Accessed on November 13, 2002.

CIMS. (2002). "Computer Integrated Management System." Online at http://www.bre.polyu.edu.hk. Accessed on November 12, 2002.

Clark, P. (1972). *Action Research and Organizational Change*, Harper & Row, London.

Corey, M., Abbey, M.; and Abramsom, I. (1998). *Oracle 8 Data Warehousing-A practical Guide to Successful Data Warehouse Analysis.* ORACLE Press.

Courage, K.; Li, I.; Hammer, J.; Ji, F.; and Yu, Q. (2004). *Feasibility Study for an Integrated Network of Data Sources*, Final report. University of Florida, Gainesville and Florida Department of Transportation, January 2004.

Cresswell, J. (1994). *Research Design: Qualitative and Quantitative Approach*, Sage Publications, London.

Domencio, J.A. (2001). *Definition of a Data Warehouse Environment in a High Education Institution*, M.Sc. thesis, Universidade Federal de Santa Catarina, Florianopolis, Brazil.

Dyché, J. (2000). *E-Data: Turning Data into Information with Data Warehousing*, Addison-Wesley, NJ.

Fisk, E.R. (2003). *Construction Project Administration*, 7th edition, Prentice Hall, NJ.

Fumo, T. G. S. (2003). *A Data Warehouse Architecture Model for the Ministry of Health in Mozambique.* MS Thesis, Department of Informatics, University of Oslo, Norway.

Galbraith, J. (1973). *Designing Complex Organizations*, Addison-Wesley, Reading, MA.

Galbraith, J.R. (1974). "Organizational Design: An Information Processing View." *Interfaces*, 4(3), 87-89.

Giachetti, R. (2002). *Information System Modeling*, Class Handouts for EIN6993 course, Florida International University, Miami, Florida.

Giaglis, G. (2001). "A Taxonomy of Business Process Modeling and Information Systems Modeling Techniques." *The International Journal of Flexible Manufacturing Systems* 13(1), 209-228.

Gibbs, A. (1997). "Focus Groups: Why use Focus Groups and not Other Methods." *Social Research Update*, 19, 17-27.

Gray, P.G., and Watson, H.J. (1998). *Decision Support in the Data Warehouse*, Prentice Hall, NJ.

Guha, S., Grover, V., Kettinger, W.J., Teng, J.T.C. (1997). "Business Process Change and Organizational Performance." *Journal of MIS*, 14(1), 119-154.

Harmon, C. (1998). "Safeguarding the Data Warehouse." *Computer Fraud and Security*, 6, 16-19.

Hastak, M. (1994). *Decision Support Systems for Project Cost Control-Strategy and Planning*, Ph.D. Thesis, Purdue University, USA.

Howard, H.C. (1991). "Linking Design Data with Knowledge-based Construction Systems." *Proceedings of the CIFE Spring Symposium*, Stanford University, 19-31.

Hult, M., and Lennung, S.A. (1980). "Towards a Definition of Action Research: A Note and Bibliography." *Journal of Management Studies*, 5, 241-250.

Hwang, H-G, Ku, C-Y, Yen, D.C., Cheng, C-C. (2002). "Critical Factors Influencing the Adoption of Data Warehouse Technology: A Study of the Banking Industry in Taiwan." *Decision Support Systems,* 1039, 1-21.

Inmon, W.H. (2000). *Building the Data Warehouse: Getting Started*, Online at http://www.billinmon.com/library/whiteprs/earlywp/ttswdev.pdf, Accessed on March 02, 2003.

Inmon, W.H., and Hackathorn, R.D. (1994). *Using the Data Warehouse*, Wiley, NY.

Jacobsen, K; Eastman, C; and Jeng, T.S. (1997). "Information Management in Creative Engineering Design and Capabilities of Database Transactions." *Automation in Construction*, 6(7), 55-69.

Jain R. K. (1997). "Matrices of Organization Effectiveness." *Journal of Management in Engineering*, 13(2), 40-45.

Kahkonen, K.E. (1994). "Interactive Decision Support System for Building Construction Scheduling." *Journal of Computing in Civil Engineering*, 8 (4), 519-535.

Kennedy, P.W. (1994). "Information Processing and Organization Design." *Journal of Economic Behavior and Organization*, 25, 37-51.

Kim, H-W. (2001). "Modeling Inter- and Intra-Organizational Coordination in Electronic Commerce Deployment." *Information Technology and Management*, 2, 335-354.

Kimball, R. (1997). *The Data Warehouse Toolkit*, Thomson International, NJ.

Kimball, R., and Caserta, J. (2004). *The Data Warehouse ETL Toolkit*, Wiley, NY.

Kimball, R. and Ross, M. (2002). *The Data Warehouse Toolkit*, 2nd edition, Wiley, NY.

Kimball, R., Reeves, L., Ross, M., and Thornthwaite, W. (1998). *The Data Warehouse Lifecycle Toolkit: Expert Methods for Design, Developing and Deploying Data Warehouses*. John Wiley & Sons, NY.

Kosanke, K., Vernadat, F., and Zelm, M. (1999). "CIMOSA: Enterprise Engineering and Integration." *Computers in Industry*, 40, 83-97.

Krishna, S.J. (2004). *Data Warehousing: Design and Development Perspectives*, ICFAI Books, India.

Kroenke, D., and Hatch, R. (1994). *Management of Information Systems*, 3rd edition, McGraw Hill, Singapore.

Kubr, M. (1986). *Management Consulting: A Guide to the Profession*, 2nd edition, Geneva: International Labor Office.

Kumaraswamy, M.M., and Dissanayaka, S.M. (2001). "Developing a Decision Support System for Building Project Procurement." *Building and Environment*, 36, 337-349.

Lee, J.K., and Lee, H.S. (2003). "Principles and Strategies for Applying Data Warehouse Technology to Construction Industry." *Architectural Research*, 5 (1), 61-68.

Levene, M., and Loizou, G. (2003). "Why is the Snowflake Schema a Good Data Warehouse Design?." *Information Systems*, 28(3), 225-240.

Love, P.E.D., and Irani, Z. (2001). "Evaluation of IT Costs in Construction." *Automation in Construction*, 10 (6), 649-658.

Love, P.E.D., Irani, Z., Edwards, D.J. (2005). "Researching the Investment of Information Technology in Construction: An Examination of Evaluation Practices." *Automation in Construction*, 14(4), 569-582.

Lippitt, G. and R. Lippit. (1978) *The Consulting Process In Action*, University Associates, San Diego, CA.

Lykins, H., Rose, S., Scott, P.C. (1998). *An Information Model for Integrating Product Development Processes*, http://www.software.org/pub/externalpapers/9804-3.html (accessed on online in October 2004).

Mak, S. (2001). "A Model of Information Management for Construction using Information Technology." *Automation in Construction*, 10, 257-263.

Mattison, R. (1996). *Data Warehousing: Strategies, Technologies, and Techniques*, McGraw-Hill, New York.

Miami-Dade County. (2005). *Miami-Dade Transit: Facts at a Glance*, http://www.miamidade.gov/transit/ (accessed online on May 21, 2005).

MDT Employee Relations Department. (2005). *Job Description/Pay Plans*, Online at http://www.miamidade.gov/emprel/pay_plan/C_PAYRANGE.htm, Accessed on June 20, 2005.

Mintzberg, H. (11993). *Structures in Five: Designing Effective Organizations*. Prentice Hall, NJ.

Mitropoulos, P., and Taum, C.B. (2000). "Management-Drive Integration." *Journal of Management in Engineering*, 16(1), 48-58.

ML Payton Consultants. (2002). "Use of Enterprise Resource Planning in the Construction Industry –Summary of Findings." *On-line at* http://www.mlpayton.com/pages/summaries.html, Accessed on November 12, 2002.

Naoum, S.G. (2001). *Dissertation Research and Writing for Construction Students*, Butterworth Heinemann, UK.

Ng, S.T., Palaneeswaran, E., and Kumaraswamy, M.M. (2003). "Web-based Centralized Multiclient Cooperative Contractor Registration System." *Journal of Computing in Civil Engineering*, 17(1), 28-37.

Nosek, J.T. (1989). "Organization Design Strategies to Enhance Information Resource Management." *Information and Management*, 16, 81-91.

nSpin Case Study. (2002). "Flexible Access to Enterprise Data and Decision Support." Online at http://www.nspin.com/industries/cases.html , accessed on October 21, 2002.

Ogunlana, S. O., Li, H., Sukhera, F.A. (2003). "System Dynamics Approach to Exploring Performance Enhancement in a Construction Organization." *Journal of Construction Engineering and Management*, 129(5), 528-536.

Oliver, D. W., Kelliher, T.B., Keegan, J.G. (1997). *Engineering Complex Systems with Models and Objects*, McGraw-Hill, New York, NY.

Palaneeswaran, E., and Kumaraswamy, M. (2005). "Web-based Client Advisory Decision Support System for Design-Builder Prequalification." *Journal of Computing in Civil Engineering*, 19(1), 69-82.

Palisades Group. (2004). *Information Technology and Intelligent Transportation Systems: 2003-2008 Strategic Plan*, Miami-Dade Transit, Miami, Florida.

Park, Y-T. (2005). "An Empirical Investigation of the Effects of Data Warehousing on Decision Performance." *Information and Management,* Corrected proof available at http://www.sciencedirect.com/science?_ob=ArticleListURL&_method=list&_ArticleListID=307274635&_sort=d&_st=4&_acct=C000054271&_version=1&_urlVersion=0&_userid=2139759&md5=f238b3c3d7607d8e73a921a1739194f9, Accessed on June 12, 2005.

Paulson B., and Kim K. (1999). "Decentralized Decision Making in Project Scheduling and Control." Online at http://www.stanford.edu/group/CIFE/cifeprojectsummary.html, Accessed on January 12, 2005.

Polevoy, A. (1999). "Data Warehousing: A Tool for Facilitating Assessment." *Proceedings of 29th ASEE/IEEE Frontiers in Education Conference*, November 10-13, San Juan, Puerto Rico, 11b1 7-11.

Powell, R.A., Single, H.M. (1996). "Focus Groups." *International Journal of Quality in Health Care*, 8(5), 499-504.

Prolog Manager. (2002). "Prolog Manager: Power Application for Total Project Control." Online at http://www.mps.com/products/PM/index.asp, Accessed on November 12, 2002.

Rachmat, S. (2000). *Australian Data Warehousing Practice*, M. Comp. thesis, School of Information Management and Systems, Monash University, Australia.

Reason, P., and Bradbury, H. (2001). *Handbook of Action Research: Participative Inquiry & Practice*. Sage Publications, London.

Rischmoller, L., and Alarcon, L.F. (2002). "4D-PS: Putting an IT New Work Process into Effect." *Proceedings of the CIB W78 Conference*, June 12-14, 2002, 1-6.

Robey, D., and Sales, C. (1994). *Designing Organizations*, McGraw-Hill Irwin, NJ.

Robbins, S. P. (2003). *Organizational Behavior (10th edition)*, Prentice Hall, New Jersey

Russel, J.S., and Skibniewski, M.J. (1990). "QUALIFIER-2: Knowledge-based System for Contractor Prequalification." *Journal of Construction Engineering and Management*, 116(1), 157-171.

Shen, Q., Chung, J.K.H., Li, H. and Shen, L. (2004). "A Group Support System for Improving Value Management Studies in Construction." *Automation in Construction*, 13(13), 209-224.

Shi, J. and Halpin, D. "Construction Enterprise Resource Planning (Construct-ERP)." *On-line*. http://www.eng.usf.edu/nsf/conference/scalable/scalable_enterprises/Shi_IIT.ppt, Accessed on November 12, 2002.

Shin, B. (2001). "A Case of Data Warehousing Project Management." *Information and Management*, 1975, 1-12.

Soibelman, L. (2000). "Construction Knowledge Generation and Dissemination." Online at http://www.ce.berkeley.edu/~tommelein/CEMworkshop/Soibelman.pdf, Accessed on January 12, 2005.

Soibelman, L., and Kim, H. (2002). "Data Preparation Process for Construction Knowledge Generation through Knowledge Discovery in Databases." *Journal of Computing in Civil Engineering*, 16(1), 39-48.

Songer, A.D., Ibbs, C.W., Garrett, J.H., Napier, T.R., and Stumpf, A.L. (1992). "Knowledge-based Advisory System for Public-Sector Design-Build." *Journal of Computing in Civil Engineering*, 6(4), 456-471.

SPACE. (2002). "Simultaneous Prototyping for An integrated Construction Environment." Online at http://www.surveying.salford.ac.uk/aic/space.htm. Accessed on November 12, 2002.

Stewart, D. and Michael, K. (1993). *Secondary Research: Information Sources and Methods*, 2nd edition, Sage Publications, London.

Subramanium, A., Smith, D. L., Nelson, A.C., Campbell, J.F., and Bird, D.A. (1997). "Strategic Planning for Data Warehousing: A Case Study." *Information and Management*, 33, 99-113.

Summer, E. and Ali, D.L. (1996). "A Practical Guide for Implementing Data Warehousing." *Proceedings of the 19th International Conference on Computers and Industrial Engineering*, Elsevier Science, UK, 307-310.

Susman, G. and Evered, R. (1978) "An Assessment of the Scientific Merits of Action Research." *Administrative Science Quarterly*, 23(4), 582-603.

Tang, S.L., Ahmad, I.U., Ahmed, S.M., Ming, L. (2004). *Quantitative Techniques for Decision Making in Construction*, Hong Kong University Press, Hong Kong.

Tang, S.L., Poon, S.W., Ahmed, S.M., and Wong, F.K.W. (2003). *Modern Construction Project Management*, Hong Kong University Press, Hong Kong.

Tenah, K.A. (1994). "Management of Information in Organizations and Routing." *Journal of Construction Engineering and Management*, 110(1), 101-118.

Trochim, W.M.K. (2001). *The Research Methods Knowledge Base*, 2nd edition, Atomic Dog Publishing, Tampa.

Tserng, H.P., and Lin, P.H. (2002). "An Accelerated Subcontracting and Procuring Model for Construction Projects." *Automation in Construction,* 11(1) 105-125.

Tushman, M.L., Nadler, D.A. (1978). "Information Processing as an Integrating Concept in Organizational Design." *Academy of Management Review*, 3, 613-624.

Vanegas, J., Chinowsky, P. (1996). "Simulation Technologies for Planning Heavy Construction." *Journal of Computing in Civil Engineering*, 10(2), 28-37.

Vassiliadis, P, Quix, C., Vassiliou, Y., and Jarke, M. (2001). "Data Warehouse Process Management." *Information Systems,* 26, 205-236.

Vernadat, F. (1993). "Integrated Model for Requirement Definition and Design Specification of CIM Systems." International Conference on Computer-Integrated Manufacturing, Bejing, May 12-14, 651-659.

VTT. (2002). "Development of 4D CAD Systems at VTT." Online at http://www.vtt.fi. Accessed on November 12, 2002.

Watson, H.J. (2001). "Designing Data Warehouses." *Data & Knowledge Engineering*, 31, 279-301.

Watson, H.J.; Goodhue, D.L.; and Wixom, B.H. (2001). "The Benefits of Data Warehousing: Why Some Organizations Realize Exceptional Payoffs." *Information and Management*, 1-12.

White, C.J. (2000). *An Analysis-led Approach to Data Warehousing Design and Development*, Database Associates, MD.

Whitten, J.L., Bentley, L.D., and Dittman, K.C. (1998). *Systems Analysis and Design Methods*, 5th edition, McGraw-Hill Irwin, NY.

Wood-Harper, T. (1989). *Comparison of Information Systems Definition Methodologies: An Action Research Multiview Perspective*, Ph.D. Thesis, University of East Anglia.

Yang, J.B., and Yau, N.J. (1996). "Application of Cased-based Reasoning in Construction Engineering and Management." *Proceedings of the Third Congress held in Conjunction with A/E/C Systems*, Computing in Civil Engineering, American Society of Civil Engineers, New York, 663-669.

Yates, J.K. (1993). "Construction Decision Support System for Delay Analysis." *Journal of Construction Engineering and Management*, ASCE, 119(2), 226-244.

Yin, R. (1994). *Case Study Research: Design and Methods*, Sage Publications, London.

Zelm, M. (1995). *CIMOSA: A Primer on Key Concepts, Purpose and Business Value*, CIMOSA Association, Stuttgart

Zhiliang, M., Wong, K.D., Heng, L., Jun, J. (2005). "Utilizing Exchanged Documents in Construction Projects for Decision Support Based on Data Warehousing Technique." *Automation in Construction*, 14(3), 405-412.

Zikmund, W. (1997). *Business Research Methods*, Dryden Press, UK.

Questionnaire Survey # 1

ARE YOUR CONSTRUCTION DATA HELPFUL IN MAKING DECISIONS?

Introduction and Objectives: Management of construction projects involves handling of a vast amount of data in the process of project planning, design, scheduling, estimating, cost control and quality control. If properly stored and managed, data can be used for effective decision making.

The objective of this questionnaire survey is to assess the data needs of **construction owner organizations** (organizations who own and manage the facility) and level of their database/information systems use. This information will be used in a research project to develop a decision making framework for construction executives. The proposed decision making framework will aid decision makers in owner organizations to quickly analyze existing data so that predictions and forecasts can be made with reasonable accuracy.

The questionnaire is designed for organization executives (i.e., CEO/President, VPs, Design/Construction Managers, Unit heads, IT/IS managers) who are involved in the decision making process and have some knowledge about the information systems use in their organization.

The questionnaire is divided into four sections and should take about **10 minutes** to complete. Your contribution towards this study is greatly appreciated, as it will add significantly to the value of the research. All information provided through this questionnaire will eventually be compiled and presented as part of a report. **YOUR RESPONSES WILL BE KEPT SECURELY AND WILL REMAIN CONFIDENTIAL**. No individual or organization will be identified in the report.

If you have any questions or require further information, please e-mail xxx@xxxx.xxx

THANK YOU VERY MUCH FOR YOUR KIND COOPERATION AND TIME

1. What describes your organization type? *(please select one)*

   ☐ Public agency or department   ☐ Private company
   ☐ Other (please specify):

2. What describes your main job function within the organization? *(Please select one)*:

   ☐ Executive (CEO, VP, Director)   ☐ Project Planner
   ☐ CIO (Systems/IT manager)   ☐ Project Engineer
   ☐ Construction Chief   ☐ Design Engineer
   ☐ Design Chief   ☐ Site Manager
   ☐ Project Manager   ☐ Other (please specify):

3. Type of construction projects your organization is involved with: *(you may select more than one)*

   ☐ Infrastructure (roads, bridges, etc.)   ☐ Commercial buildings
   ☐ Institutional buildings (schools, hospitals)   ☐ Residential (single, multi family, etc)
   ☐ Other (please specify):

4. Number of projects undertaken in the last five years: *(please select one)*

   ☐ < 10   ☐ 10 - 30   ☐ 30 - 50   ☐ >50

5. Number of employees: *(please specify)*

   Technical: Administrative:

6. Organization's annual construction expenditure (US$): *(please select one)*

   ☐ Less than $25 million   ☐ Between $25 million and $50 million
   ☐ Between $50 million and $100 million   ☐ More than $100 million

7. How does your organization store and manage project data? *(please select on the basis of your assessment of the extent of use of each method list below)*

   Manual (Storing data sheets including computer printouts in paper files and folders)
   Computerized (Storing data using a database management system such as Access, Excel, etc)

Manual / Computerized (%)	0/100	25/75	50/50	75/25	100/0
	☐	☐	☐	☐	☐

8. What type of computerized database management system (DBMS) is used in your organization to store and manage project data? *(you may select more than one)*

   (Answer this question only if your organization uses any computerized DBMS)

   ☐ Oracle   ☐ MS Access
   ☐ PeopleSoft, SAP, BAAN   ☐ Fox Pro, DB2, DB3, DB4
   ☐ Spreadsheet (Excel, Lotus etc.)   ☐ Other (please specify):

9. Please indicate your level of satisfaction with the existing database management system using the following criteria:

	Very Satisfied	Satisfied	Neutral	Dissatisfied	Very Dissatisfied
Ease in data access	☐	☐	☐	☐	☐
Data quality	☐	☐	☐	☐	☐
Productivity improvement	☐	☐	☐	☐	☐
Quality of reports	☐	☐	☐	☐	☐
Support for decision making	☐	☐	☐	☐	☐
Cost of operation	☐	☐	☐	☐	☐

10. Does your existing information system(s) provide any of the following functionalities?

	Yes	No
Data integration (e.g. integrate data from different processes or information sources)	☐	☐
Generate data trends over a period of time (e.g. productivity trend over time)	☐	☐
Generate data summaries at various levels (e.g. for executive management, middle management etc.)	☐	☐
Perform "What-If" analysis (examine the effect of changing one variable over the other, e.g. Examine increase/decrease in productivity by hiring part time labor)	☐	☐

11. Please indicate your level of agreement regarding the following activities in your organization?

	Strongly Agree	Agree	Neither/nor	Disagree	Strongly Disagree
Use of existing IS to make everyday decisions	☐	☐	☐	☐	☐
Use of IS for short and long term planning (e.g. using trend analysis or other such techniques)	☐	☐	☐	☐	☐
Use of IS to prepare summarized reports for executive decisions	☐	☐	☐	☐	☐

12. If you would like to improve your existing information system (IS), which solution you will prefer? *(please select one)*

☐ Hiring additional staff to more effectively use the existing IS
☐ Purchasing a new and enhanced IS
☐ Both hiring additional staff and purchasing a new IS
☐ Other (please specify):

Data warehousing is a fairly recent development in the field of database and information systems (IS). A data warehouse is a specialized database created by combining data from multiple existing databases for purposes of analysis. With the appropriate user-friendly query tools, users can experiment with different views of the data, thus gaining a better understanding of the situation in order to make improved decisions. Data warehouse provides direct answer to many questions or queries asked at the executive level, such as the *what-if* and *what-next* type questions, data trends over a period of time and summarized reports. The cost and implementation time of data warehousing varies depending on organization size and extent of automation.

13. Do you think data warehousing is needed in your organization? ☐ Yes ☐ No

14. If No, what are the main reasons? *(you may select more than one)*

    ☐ Cost is high as compared to its benefits
    ☐ Implementation time is long
    ☐ Size of organization is small
    ☐ Trend analysis is not important for short and long term planning
    ☐ Computers are unreliable
    ☐ Data ownership problem (data is owned by different departments which do not want to share it)
    ☐ Executive management is not interested to invest in new information technology solutions
    ☐ Other (please specify):

15. When your organization decides to implement a new information technology solution (such as software application, hardware, networking tools etc.) how does that investment is justified within the organization?

    ☐ By conducting a cost/benefit or value analysis
    ☐ By comparing competitive advantage of the new system with the current ones
    ☐ Other (please specify):

4: Contact Information (Optional)

1. Organization name:

2. Organization location:
    City   State

3. Your name:

4. Phone no.:

5. E-mail address:

6. Please indicate if you would like to be acknowledged in the report. Your specific response to this questionnaire will not be mentioned with your identity: ☐ Yes ☐ No

7. Would you like to have a copy of the report? ☐ Yes   ☐ No

DATA WAREHOUSE DESIGN SCHEMAS

B.1 Star Schemas for Various Data Marts

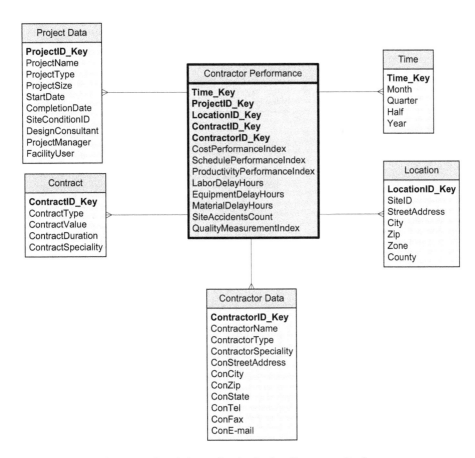

Figure B1: Star Schema for Analyzing Contractor Performance

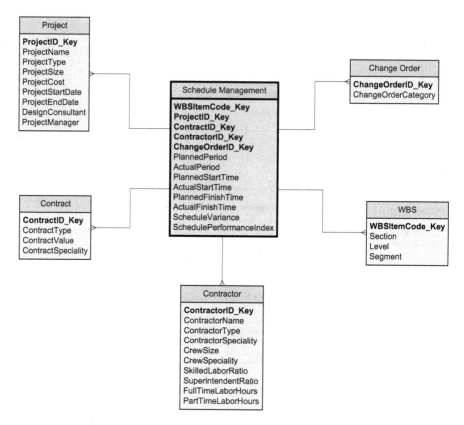

Figure B2: Star Schema for Project Schedule Management

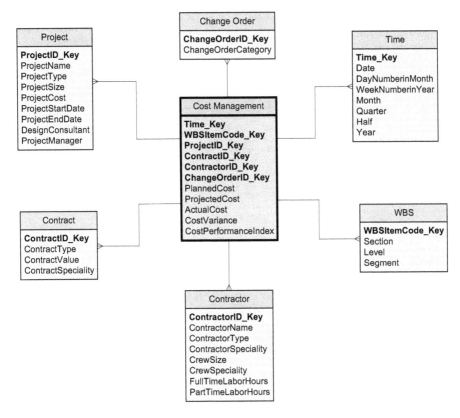

Figure B3: Star Schema for Project Cost Management

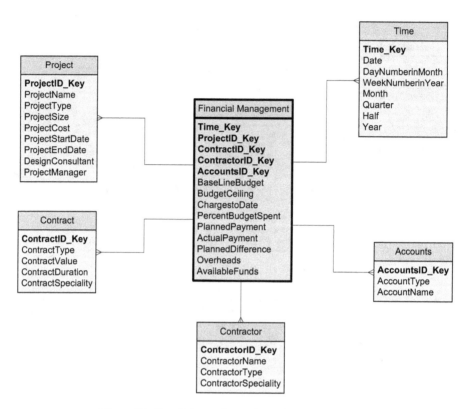

Figure B4: Star Schema for Project Financial Management

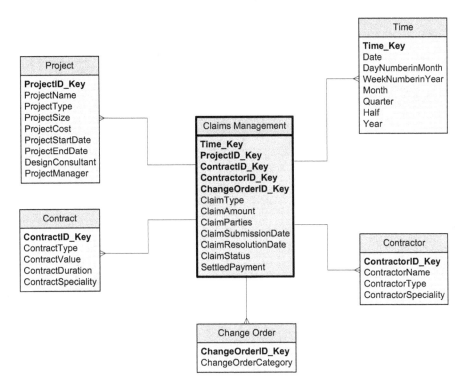

Figure B5: Star Schema for Project Claims Management

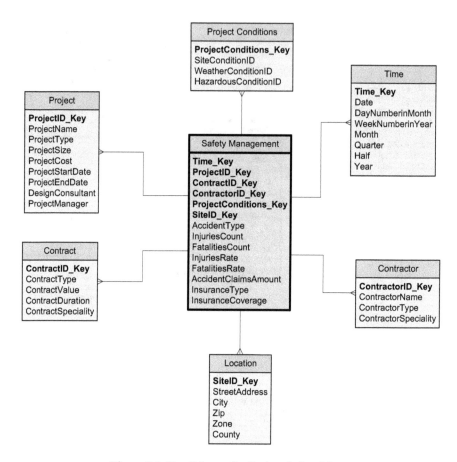

Figure B6: Star Schema for Project Safety Management

B.2 Metadata of Facts and Dimensions

Table B1: Metadata of Facts

No.	Data Mart	Measure	Domain	Data Type
1.	Project Performance	Planned Cost, Actual Cost, Cost Variance	$0.01 - $9,999,999.99	Currency (9,2)
		Planned Duration, Actual Duration, Schedule Variance, Cost Performance Index, Schedule Performance Index	0 – 9,999.99	Numeric (6,2)
		Percent Completion	0.01 – 99.99	Numeric (4.2)
		Accidents Count, Fatality Count	1 – 9,999	Numeric (4,0)
		Cost Changes due to Change Orders	$0.01 - $9,999,999.99	Currency (9,2)
		Project Claims Count	1 – 9,999	Numeric (4,0)
		Claims Amount	$0.01 - $9,999,999.99	Currency (9,2)
2.	Productivity Information	Measured Productivity	0.01 – 9,999.99	Numeric (6,2)
		Productivity Performance Index	0.01 – 9.999	Numeric (1,3)
		Equipment Delay Hours, Material Delay Hours, Labor Delay Hours, Miscellaneous Delay Hours	0.01 – 9,999.99	Numeric (6,2)
		Change Orders Count, RFIs Count	1 – 9,999	Numeric (4,0)
3.	Contractor Performance	Cost Performance Index, Schedule Performance Index, Productivity Performance Index	0.01 – 9,999.99	Numeric (6,2)
		Equipment Delay Hours, Material Delay Hours, Labor Delay Hours	0.01 – 9,999.99	Numeric (6,2)
		Site Accidents Count	1 – 9,999	Numeric (4,0)
		Quality Measurement Index	0.01 – 9,999.99	Numeric (6,2)
4.	Schedule Management	Planned Period, Actual Period	0 – 9,999	Numeric (4,0)
		Planned Start Time, Actual Start Time, Planned Finish Time, Actual Finish Time	Valid Dates	Date/Time
		Schedule Variance, Schedule Performance Index	0 – 9,999.99	Numeric (6,2)
5.	Cost Management	Planned Cost, Projected Cost, Actual Cost	$0.01 - $9,999,999.99	Currency (9,2)
		Cost Variance, Cost Performance Index	0 – 9,999.99	Numeric (6,2)

No.	Data Mart	Measure	Domain	Data Type
6.	Financial Management	Base Line Budget, Budget Ceiling, Charges to Date, Planned Payment, Actual Payment, Planned Difference, Overheads, Available Funds	$0.01 - $999,999,999.99	Currency (11,2)
		Percent Budget Spent	0.01 – 99.99	Numeric (4.2)
		Claim Amount	$0.01 - $9,999,999.99	Currency (9,2)
		Claim Parties, Claim Status	---	Character (50)
		Claim Submission Date, Claim Resolution Date	Valid Dates	Numeric
		Settled Payment	$0.01 - $9,999,999.99	Currency (9,2)
8.	Safety Management	Accident Type	---	Character (50)
		Injuries Count, Fatalities Count	1 – 9,999	Numeric (4,0)
		Injuries Rate, Fatalities Rate	0 – 9,999.99	Numeric (6,2)
		Accident Claim Amount	$0.01 - $9,999,999.99	Currency (9,2)
		Insurance Type	---	Character (20)
		Insurance Coverage	$0.01 - $9,999,999.99	Currency (9,2)

Table B2: Metadata of Dimensions

No.	Dimension	Attributes	Domain	Data Type
1.	Time	Date	Valid Dates	Date/Time
		Day of Week	1 – 7	Character (1)
		Day Number in Month	1 - 31	Character (2)
		Day Number Overall	1 – 9,999	Character (4)
		Week Number in Year	1 - 52	Character (2)
		Week Number Overall	1 – 9,999	Character (4)
		Month	January – December	Date/Time
		Month Number Overall	1 – 9,999	Character (4)
		Quarter	1 – 4	Character (1)
		Fiscal Quarter	1 – 999	Character (3)
		Half	1 – 999	Character (3)
		Year	1990 - 2050	Date/Time
		Fiscal Year	1 - 99	Character (2)
		Holiday Flag	--	Character (10)
2.	Location	Site ID	--	Character (8)
		Street Address	--	Character (30)
		City	Valid City Name	Character (15)
		Zip	Valid Zip Code	Numeric (5)
		Zone	Valid Zone Code	Numeric (4)
		County	Miami-Dade	Character (10)
3.	Project	Project ID	---	Character (12)
		Project Name	---	Character (20)
		Project Type	Bridge, Road Work, Facility	Character (10)
		Project Size	0 – 9,999.99	Numeric (6,2)
		Project Cost	$0.01 - $9,999,999.99	Currency (9,2)
		Project Start Date, Project End Date	Valid Dates	Date/Time
		Design Consultant, Project Manager	---	Character (25)
4.	Contract	Contract ID	--	Character (8)
		Contract Type	Valid Type from Available Pool	Character (15)
		Contract Value	$0.01 - $9,999,999.99	Currency (9,2)
		Contract Duration	1 – 999	Character (3)
		Contract Specialty	--	Character (15)

Table B2: Metadata of Dimensions (Continue)

No.	Dimension	Attributes	Domain	Data Type
5.	Contractor	Contractor ID	--	Character (8)
		Contractor Name	--	Character (30)
		Contractor Type	Prime, Sub-contractor, Specialty Contractor	Character (30)
		Contractor Specialty	--	Character (30)
		Crew Size	1 – 999	Character (3)
		Crew Specialty	--	Character (30)
		Skilled Labor Ratio	0.01 – 99.99	Numeric (4.2)
		Superintendent Ratio	0.01 – 99.99	Numeric (4.2)
		Full Time Labor Hours	1 – 9,999	Character (4)
		Part Time Labor Hours	1 – 9,999	Character (4)
		Contractor's Street Address	---	Character (30)
		Contractor's City	Valid City Name	Character (15)
		Contractor's Zip Code	Valid Zip Code	Numeric (5)
		Contractor's Phone	--	Character (10)
		Contractor's Fax	--	Character (10)
		Contractor's E-mail	--	Character (30)
6.	Project Conditions	Site Conditions ID	Valid ID from Available Pool	Character (4)
		Weathers Condition ID	Valid ID from Available Pool	Character (4)
		Hazardous Conditions ID	Valid ID from Available Pool	Character (4)
7.	WBS	WBS Item Code	Valid WBS Code	Character (8)
		Section	--	Character (4)
		Level	--	Character (4)
		Segment	--	Character (4)
8.	Change Orders	Change Order ID	--	Character (8)
		Change Order Category	Valid Category from Available Pool	Character (10)
9.	Accounts	Account ID	--	Character (8)
		Account Type	Valid Category from Available Pool	Character (8)
		Account Name	--	Character (20)

B.3 Defining Multidimensional Data in *Power OLAP™*(Selected Screenshots)

Figure B7: Building Multidimensional Data Cubes

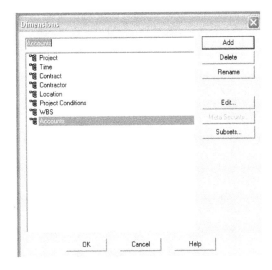

Figure B8: Adding Dimensions to Various Data Cubes

217

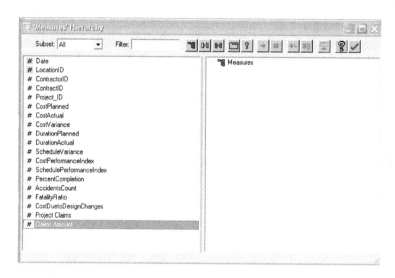

Figure B9: Defining Foreign Keys and Measures of Fact Data
for "Project Performance" Cube

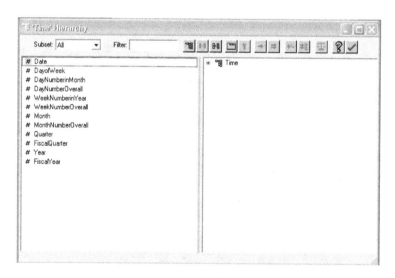

Figure B10: Defining Attributes of Dimension "Time"

218

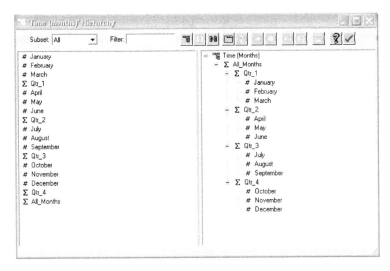

Figure B11: Adding Granularity Levels to Dimension "Time"

APPENDIX C

SAMPLE RUN

C.1 Selected Screenshots Illustrating Various OLAP Operations

C.1.1 Multidimensional Analysis

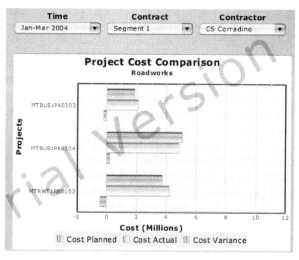

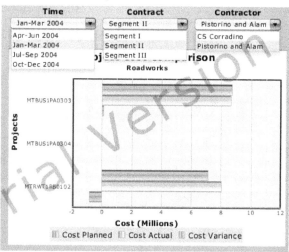

Figure C1: Project Cost Comparison using "Time", "Contract" and "Contractor" Dimensions

C.1.2 Trend Analysis

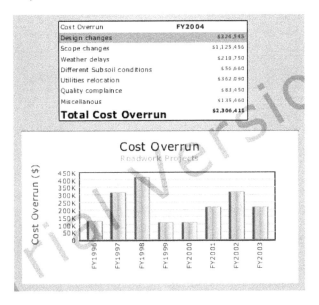

Cost Overrun	FY2004
Design changes	$324,545
Scope changes	$1,125,456
Weather delays	$218,750
Different Subsoil conditions	$56,660
Utilities relocation	$362,090
Quality complaince	$83,450
Miscellanous	$135,460
Total Cost Overrun	**$2,306,411**

Figure C2a: Trend Analysis for Cost Overrun due to "Design Changes"

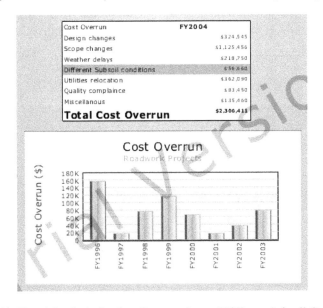

Cost Overrun	FY2004
Design changes	$324,545
Scope changes	$1,125,456
Weather delays	$218,750
Different Subsoil conditions	$56,660
Utilities relocation	$362,090
Quality complaince	$83,450
Miscellanous	$135,460
Total Cost Overrun	**$2,306,411**

Figure C2b: Trend Analysis for Cost Overrun due to "Different Subsoil Conditions"

C.1.3 Slice and Dice Analysis

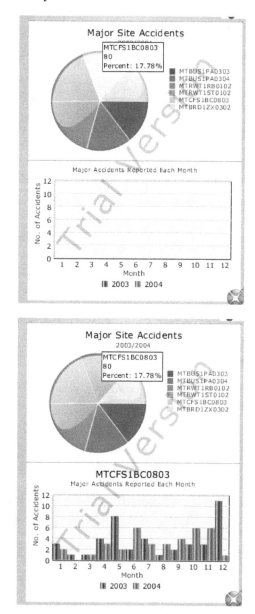

Figure C3: Slicing Site Accidents Data for Project "MTCFS1BC9803" and Dicing for Years 2003 and 2004

C.1.4 What-If Analysis

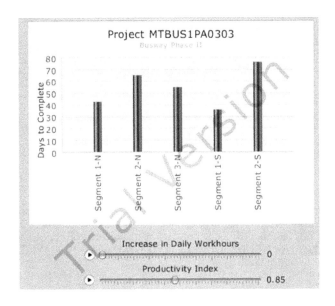

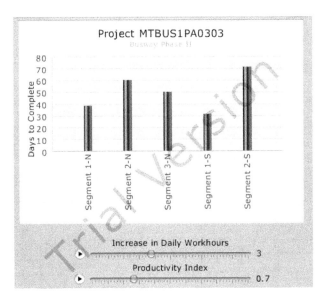

Figure C4: Forecasting Project Duration by Increasing Daily Work hours and Adjusting Productivity Index

APPENDIX D

Questionnaire Survey - 2

MEASUREMENT OF IMPACT OF PROPOSED DECISION-SUPPORT FRAMEWORK ON MIAMI-DADE TRANSIT'S (MDT) ORGANIZATIONAL PERFORMANCE

Introduction and Objectives: The objective of this questionnaire survey is to measure the impact of proposed decision-support framework on the organizational performance of Miami-Dade Transit (MDT). The impact is evaluated by both quantitative and qualitative measures. The results of this questionnaire survey will help us to analyze the improvements in different processes as a result of the proposed decision-support framework implementation.

Please answer this questionnaire in light of the demonstrated functionalities of the proposed decision-making framework. Your contribution towards this study is greatly appreciated, as it will add significantly to the value of the research. All information provided through this questionnaire will eventually be compiled and presented as part of a report. **YOUR RESPONSES WILL BE KEPT SECURELY AND WILL REMAIN CONFIDENTIAL.** No individual will be identified in the report.

1. Please indicate your level of satisfaction in terms of improvements in the following processes after using the proposed decision-support framework:

   (a) Focus on data management operations:

	No Improvements	Little Improvements	Some Improvements	Fair Improvements	Significant Improvements
	1	2	3	4	5
Ease in data access	☐	☐	☐	☐	☐
Data quality	☐	☐	☐	☐	☐
Data integration	☐	☐	☐	☐	☐
Response time to access the required data or information	☐	☐	☐	☐	☐
Generation of data trends and data summaries	☐	☐	☐	☐	☐
Quality of reports	☐	☐	☐	☐	☐

(b) Focus on decision-support operations:

	No Improvements	Little Improvements	Some Improvements	Fair Improvements	Significant Improvements
	1	2	3	4	5
Support for every day decisions	☐	☐	☐	☐	☐
Support for short and long term planning	☐	☐	☐	☐	☐
Confidence in decisions (without using "gut-feelings" or experience)	☐	☐	☐	☐	☐
Time to make decisions	☐	☐	☐	☐	☐
Personal productivity	☐	☐	☐	☐	☐
Organizational productivity	☐	☐	☐	☐	☐

2. Please evaluate the effect of decision-support framework implementation on the following factors according to your best judgment:

	Less than 10%	Between 10% - 20%	Between 20% - 30%	Between 30% - 40%	Between 40% - 50%	More than 50%
Time savings (e.g. in decision making, reports generation, etc.)	☐	☐	☐	☐	☐	☐
Operational cost savings (e.g., information gathering costs, reports generation costs, etc.)	☐	☐	☐	☐	☐	☐

Questionnaire Survey - 3

SUITABILITY OF DATA WAREHOUSING BASED DECISION-SUPPORT FRAMEWORK FOR YOUR ORGANIZATION

Introduction and Objectives: The objective of this questionnaire survey is to measure suitability of the proposed decision-support framework for possible implementation in your organization. The framework is developed using the data warehousing approach. The results of this questionnaire survey will help us to analyze the improvements in different processes as a result of the proposed decision-making framework implementation.

We are sending this questionnaire survey to you because you showed interest in the implementation of data warehousing in your organization to support decision-making operations. Please answer this questionnaire in light of the demonstrated functionalities of the proposed decision-support framework as shown in the attached notes. Your contribution towards this study is greatly appreciated, as it will add significantly to the value of the research. All information provided through this questionnaire will eventually be compiled and presented as part of a report. **YOUR RESPONSES WILL BE KEPT SECURELY AND WILL REMAIN CONFIDENTIAL.** No individual will be identified in the report.

If you have any questions or require further information, please e-mail xxxx@xxxxxx.xxx

---

THANK YOU VERY MUCH FOR YOUR KIND COOPERATION AND TIME

1. Do you think that the proposed decision-support framework could be successfully implemented in your organization? ☐ Yes ☐ Undecided ☐ No

   If your answer to above question is **No** or **Undecided**, what are the main reasons? *(you may select more than one)*

   ☐ Not compatible with organization needs
   ☐ Senior management may prefer a different solution
   ☐ Lack of financial resources
   ☐ Costs outweigh benefits
   ☐ Organization has already implemented or in a process of implementing a new DBMS
   ☐ Other (please specify): _____

2. Please indicate your level of satisfaction in terms of improvements in the following processes if the proposed decision-support framework is implemented in your organization:

   (a) Focus on data management operations:

	No Improvements	Little Improvements	Some Improvements	Fair Improvements	Significant Improvements
	1	2	3	4	5
Ease in data access	☐	☐	☐	☐	☐
Data quality	☐	☐	☐	☐	☐
Data integration	☐	☐	☐	☐	☐
Response time to access the required data or information	☐	☐	☐	☐	☐
Generation of data trends and data summaries	☐	☐	☐	☐	☐
Quality of reports	☐	☐	☐	☐	☐

   (b) Focus on decision-support operations:

	No Improvements	Little Improvements	Some Improvements	Fair Improvements	Significant Improvements
	1	2	3	4	5
Support for every day decisions	☐	☐	☐	☐	☐
Support for short and long term planning	☐	☐	☐	☐	☐
Confidence in decisions (without using "gut-feelings" or experience)	☐	☐	☐	☐	☐
Time to make decisions	☐	☐	☐	☐	☐
Personal productivity	☐	☐	☐	☐	☐
Organizational productivity	☐	☐	☐	☐	☐

3. Please evaluate the post-implementation effects of decision-support framework on the following factors according to your best judgment:

	Less than 10%	Between 10% - 20%	Between 20% - 30%	Between 30% - 40%	Between 40% - 50%	More than 50%
Time savings (e.g. in decision making, reports generation, etc.)	☐	☐	☐	☐	☐	☐
Operational cost savings (e.g., information gathering costs, reports generation costs, etc.)	☐	☐	☐	☐	☐	☐

4. Do you think that the organizational structure should change as proposed in the decision-support framework?

☐ Yes   ☐ Undecided   ☐ No

If **Yes**, what type of changes would you favor?

☐ Restructuring of entire organization (e.g. merger of different divisions)
☐ Restructuring of each separate division (by keeping the overall organizational structure as same)
☐ Changes in each division without any major restructuring efforts

If your answer to above question is **No** or **Undecided**, what are the main reasons?

----

5. Finally, do you foresee an actual implementation of this or similar decision-support framework in your organization?

☐ Yes   ☐ Undecided   ☐ No

If your answer to above question is **Yes**, what could be the time-frame?

☐ Within 1 year
☐ Within next 3 years
☐ Within next 5 years
☐ Within 10 years

APPENDIX F

GLOSSARY

Architecture (or Reference Architecture)
Architecture defines the functions and responsibilities for each component of an application along with technologies to implement it.

Attribute
An attribute is a descriptive property or characteristics of any entity. Attributes are the basic units of analysis in a data base or data warehouse.

Data Cube
A data cube is a representation of multidimensional data defined by a set of 'n' different dimensions. Data is organized in a data cube by calculating all its possible aggregations.

Data Flow Diagram
A data flow diagram (DFD) is a tool that depicts the flow of data through a system and the work or processing performed by that system.

Data Mart
A data mart is a data store which contains a subset or aggregation of the data of a DWH. Data marts are categorized according to functional areas depending on the domain.

Data Mining
A technique to discover data trends, patterns and associations from the data warehouse data using various mathematical or statistical techniques.

Data Warehouse
A data warehouse is typically a read-only dedicated database system created by integrating data from multiple databases and other information sources.

Data Type
The data type for an attribute defines what type of data can be stored in that attribute.

Denormalization
Denormalization is a technique to move from higher to lower normal forms of database modeling in order to speed up the database access and thus query processing.

Dimension
A dimension represents descriptive data that reflect the dimensions of an entity (or fact). For example, Time, Project, Product, etc.

Dimension Table
A dimension table is a table of data warehouse schema (i.e. Star, or Snowflake) which contains the possible values of one dimension.

Domain
The domain of an attribute defines what values an attribute can legitimately take on.

Drill Across
To change the dimensional orientation of the cube, for analyzing the data using a particular dimension level as independent variable. For example, swapping two dimensions.

Drill Down (or Roll Down)
The navigation amongst levels of data ranging form higher summary level to lower summary level (summarized to detailed data). The drilling paths are defined by the hierarchies within dimensions.

Entity-Relationship Model (ER model)
ER modeling is a technique to visualize relational data models. Classes of objects are identified and modeled as entities. Entities have properties that are either attributes that directly describe entities or can be relationships that model facts about entities.

Fact
Facts represent quantitative (or factual) data about a business entity/transaction.

Fact Fable
The fact table is the central table in a data warehouse schema which contains the measures of interest of the DWH and foreign keys to dimension tables.

Foreign Key
A foreign key is a primary key of one entity that is also an attribute in another entity. This attribute may or may not be a primary key in the other entity.

Functional Model
A functional (or logical) model represents what a system is or does without showing the actual implementation of the system.

Granularity
Granularity refers to the level of detail provided by a data point in the data warehouse. The more detail, the lower the level of granularity.

Hierarchy
A hierarchy defines the way in which a dimension can be grouped in a DWH. For example, a possible hierarchy for the time dimension could be year, month and day.

Hypercube
Cube with more than 3 dimensions.

IT Infrastructure
IT infrastructure specifies the hardware and software tools used in implementing a DWH.

Metadata
Metadata is defined as data about data. It is like a card index describing how information is structured within the data mart, i.e. it defines field names, their data type, field sizes, data format, any validation rules and other such attributes.

Middleware
Middleware is connectivity software that consists of a set of enabling services that let multiple processes run on one or more machines to interact across a network.

Normalization
Normalization is the process of reorganizing data to minimize data redundancy. It usually involves dividing a database table into several tables and defining a relationship between the tables. Without normalization, redundant data in a database can create inconsistencies and update anomalies can occur during deletion and insertion processes.

Primary Key
A primary key is an attribute, or a group of attributes, that assumes a unique value for each entity instance.

Slice and Dice
The term *slice and dice* refers to the idea that the user can extract portions of the aggregated data and examine it in detail according to the dimensions of interest.

OLAP (Online Analytical Processing)
OLAP is the name given to the dynamic data analysis required to create, manipulate, animate and synthesize information from a DWH. This includes the ability to discern new or unanticipated relationships between variables, to create an unlimited number of dimensions and to specify cross-dimensional conditions and expressions.

OLTP (Online Transaction Processing)
Transaction-oriented work with operational systems.

Return-on-Investment (ROI)
Return-on-Investment compares the lifetime profitability of alternative solutions or projects.

Roll Up
Roll up is the inverse operation to a drill down; more precisely the navigation amongst levels of data ranging from lower summary level to higher summary level.

Snowflake Schema

The snowflake schema is a variation of the star structure, in which all dimensional information is stored in the third normal form (i.e., dimension tables have subdimension tables to avoid dependency of non-key attributes), while keeping fact table structure the same.

Star Schema

The star schema is the simplest database structure containing a fact table in the center which is surrounded by the dimension tables. The star schema uses denormalized data to provide fast response times, allowing database optimizers to work with simple database structures in order to yield better execution plans.

Unified Modeling Language (UML)

The UML is a set of modeling conventions that is used to specify or describe a software system in terms of objects. UML itself is not a programming language, it is only a notation or syntax to define an object oriented model.

(Sources: Kimball and Ross, 2004; Fumo, 2003; Diché, 2000; Gray and Watson, 1998)

APPENDIX G

LIST OF ACRONYMS

AI	Artificial Intelligence
ANOVA	Analysis of Variance
API	Application Interface Protocol
BI	Business Intelligence
C2C	Center-to-Center
CD	Construction Division (of Miami-Dade Transit)
CIMOSA	Computer Integrated Manufacturing Open Systems Architecture
CORBA	Common Object Request Broker Interface
CMS	Construction Management Services
DB	Database
DBMS	Database Management System
DEE	Design and Engineering Division (of Miami-Dade Transit)
DFD	Data Flow Diagram
DM	Data Mart
DMi	Data Mining
DOT	Department of Transportation
DSS	Decision-Support System
DWH	Data Warehouse
DWS	Data Warehouse System
EIS	Executive Information System
ER	Entity-Relation

EVA	Earned Value Analysis
GDSS	Group Decision-Support System
GSS	Group Support System
GUI	Graphical User Interface
GUI	Graphical User Interface
IDEF	Integrated Definition for Function Modeling
IS	Information System
ISP	Internet Service Provider
IT	Information Technology
KDD	Knowledge Data Discovery
KM	Knowledge Management
LAN	Local Area Network
MDT	Miami-Dade Transit
MIS	Information Management System
MOLAP	Multidimensional Online Analytical Processing
MS	Microsoft
NPV	Net Present Value
ODBC	Open Database Connectivity
OLAP	Online Analytical Processing
OLTP	Online Transaction Processing
OO	Object Oriented
PCD	Project Control Division (of Miami-Dade Transit)
QA	Quality Assurance

QM	Quality Management
RAD	Rapid Application Design
RAID	Redundant Array of Inexpensive Disks
ROI	Return on Investment
ROLAP	Relational Online Analytical Processing
SDLC	Systems Development Life Cycle
SQL	Structured Query Language
UML	Unified Modeling Language
VBA	Visual Basic for Applications
WAN	Wide Area Network
WWW	World Wide Web

DISCLAIMER

The opinions and recommendations expressed in this book are the author's personal opinions based on the collected data and do not necessarily represent Miami-Dade Transit (MDT) and other participating organizations' official position. The data warehousing implementation results shown in Appendix C and elsewhere are for demonstration purposes only and do not reflect the actual project data and/or project scenarios.

The author is not liable to anyone for any loss or damage caused by any error or omission in the research results, whether such error or omission is the result of negligence or any other cause. All such liability is disclaimed.